ISBN 0-8373-2998-1

C-2998   CAREER EXAMINATION SERIES

This is your
PASSBOOK® for...

Food Inspector Trainee

Test Preparation Study Guide

Questions & Answers

NATIONAL LEARNING CORPORATION

PASSBOOK®

NOTICE

PASSBOOK® SERIES

THE *PASSBOOK® SERIES* has been created to prepare applicants and candidates for the ultimate academic battlefield — the examination room.

At some time in our lives, each and every one of us may be required to take an examination — for validation, matriculation, admission, qualification, registration, certification, or licensure.

Based on the assumption that every applicant or candidate has met the basic formal educational standards, has taken the required number of courses, and read the necessary texts, the *PASSBOOK® SERIES* furnishes the one special preparation which may assure passing with confidence, instead of failing with insecurity. Examination questions — together with answers — are furnished as the basic vehicle for study so that the mysteries of the examination and its compounding difficulties may be eliminated or diminished by a sure method.

This book is meant to help you pass your examination provided that you qualify and are serious in your objective.

The entire field is reviewed through the huge store of content information which is succinctly presented through a provocative and challenging approach — the question-and-answer method.

A climate of success is established by furnishing the correct answers at the end of each test.

You soon learn to recognize types of questions, forms of questions, and patterns of questioning. You may even begin to anticipate expected outcomes.

You perceive that many questions are repeated or adapted so that you can gain acute insights, which may enable you to score many sure points.

You learn how to confront new questions, or types of questions, and to attack them confidently and work out the correct answers.

You note objectives and emphases, and recognize pitfalls and dangers, so that you may make positive educational adjustments.

Moreover, you are kept fully informed in relation to new concepts, methods, practices, and directions in the field.

You discover that you are actually taking the examination all the time: you are preparing for the examination by "taking" an examination, not by reading extraneous and/or supererogatory textbooks.

In short, this PASSBOOK®, used directedly, should be an important factor in helping you to pass your test.

FOOD INSPECTOR TRAINEE

DUTIES: As a **Food Inspector Trainee**, you would be in training for the position of Food Inspector and would work under close supervision and undergo a formal training program. The training provides the experience necessary to conduct inspections at food manufacturers, processors, distributors and retailers; and to report violations of standards as they pertain to wholesomeness, labeling and advertising of food products. Upon successful completion of the one year training program, you would be advanced to Food Inspector without further examination.

SCOPE OF EXAMINATION: There will be a **written test** which you must pass in order to be considered for appointment. The **written test** is designed to test for knowledge, skills and/or abilities in such areas as:

1. **Evaluating Information and Evidence** - These questions test for the ability to evaluate and draw conclusions from information and evidence. Each question consists of a set of facts and a conclusion based on the facts. The candidate must decide if the conclusion is warranted by the facts.
2. **Food safety and food establishment sanitation** - These questions test for a knowledge of food establishment sanitation procedures. Questions will cover such sanitary operations as vermin control, sanitation of equipment, temperature control, transportation and storage monitoring, personnel hygiene, and disease control.
3. **General science concepts related to chemistry, microbiology, entomology and food science** - These questions will test for a knowledge in areas listed as they relate to subjects such as food sampling, testing, food adulteration and preservation.
4. **Preparing written material** - These questions test for the ability to present information clearly and accurately and to organize paragraphs logically and comprehensibly. For some questions, you will be given information in two or three sentences followed by four restatements of the information. You must then choose the best version. For other questions, you will be given paragraphs with their sentences out of order and then asked to choose from four suggestions the best order for the sentences.
5. **Understanding and interpreting written material** - These questions test how well you comprehend written material. You will be provided with brief reading selections and will be asked questions about the selections. All the information required to answer the questions will be presented in the selections; you will not be required to have any special knowledge relating to the subject areas of the selections.

HOW TO TAKE A TEST

I. YOU MUST PASS AN EXAMINATION

A. WHAT EVERY CANDIDATE SHOULD KNOW

Examination applicants often ask us for help in preparing for the written test. What can I study in advance? What kinds of questions will be asked? How will the test be given? How will the papers be graded?

As an applicant for a civil service examination, you may be wondering about some of these things. Our purpose here is to suggest effective methods of advance study and to describe civil service examinations.

Your chances for success on this examination can be increased if you know how to prepare. Those "pre-examination jitters" can be reduced if you know what to expect. You can even experience an adventure in good citizenship if you know why civil service exams are given.

B. WHY ARE CIVIL SERVICE EXAMINATIONS GIVEN?

Civil service examinations are important to you in two ways. As a citizen, you want public jobs filled by employees who know how to do their work. As a job seeker, you want a fair chance to compete for that job on an equal footing with other candidates. The best-known means of accomplishing this two-fold goal is the competitive examination.

Exams are widely publicized throughout the nation. They may be administered for jobs in federal, state, city, municipal, town or village governments or agencies.

Any citizen may apply, with some limitations, such as the age or residence of applicants. Your experience and education may be reviewed to see whether you meet the requirements for the particular examination. When these requirements exist, they are reasonable and applied consistently to all applicants. Thus, a competitive examination may cause you some uneasiness now, but it is your privilege and safeguard.

C. HOW ARE CIVIL SERVICE EXAMS DEVELOPED?

Examinations are carefully written by trained technicians who are specialists in the field known as "psychological measurement," in consultation with recognized authorities in the field of work that the test will cover. These experts recommend the subject matter areas or skills to be tested; only those knowledges or skills important to your success on the job are included. The most reliable books and source materials available are used as references. Together, the experts and technicians judge the difficulty level of the questions.

Test technicians know how to phrase questions so that the problem is clearly stated. Their ethics do not permit "trick" or "catch" questions. Questions may have been tried out on sample groups, or subjected to statistical analysis, to determine their usefulness.

Written tests are often used in combination with performance tests, ratings of training and experience, and oral interviews. All of these measures combine to form the best-known means of finding the right person for the right job.

II. HOW TO PASS THE WRITTEN TEST

A. NATURE OF THE EXAMINATION

To prepare intelligently for civil service examinations, you should know how they differ from school examinations you have taken. In school you were assigned certain definite pages to read or subjects to cover. The examination questions were quite detailed and usually emphasized memory. Civil service exams, on the other hand, try to discover your present ability to perform the duties of a position, plus your potentiality to learn these duties. In other words, a civil service exam attempts to predict how successful you will be. Questions cover such a broad area that they cannot be as minute and detailed as school exam questions.

In the public service similar kinds of work, or positions, are grouped together in one "class." This process is known as *position-classification*. All the positions in a class are paid according to the salary range for that class. One class title covers all of these positions, and they are all tested by the same examination.

B. FOUR BASIC STEPS

1) Study the announcement

How, then, can you know what subjects to study? Our best answer is: "Learn as much as possible about the class of positions for which you've applied." The exam will test the knowledge, skills and abilities needed to do the work.

Your most valuable source of information about the position you want is the official exam announcement. This announcement lists the training and experience qualifications. Check these standards and apply only if you come reasonably close to meeting them.

The brief description of the position in the examination announcement offers some clues to the subjects which will be tested. Think about the job itself. Review the duties in your mind. Can you perform them, or are there some in which you are rusty? Fill in the blank spots in your preparation.

Many jurisdictions preview the written test in the exam announcement by including a section called "Knowledge and Abilities Required," "Scope of the Examination," or some similar heading. Here you will find out specifically what fields will be tested.

2) Review your own background

Once you learn in general what the position is all about, and what you need to know to do the work, ask yourself which subjects you already know fairly well and which need improvement. You may wonder whether to concentrate on improving your strong areas or on building some background in your fields of weakness. When the announcement has specified "some knowledge" or "considerable knowledge," or has used adjectives like "beginning principles of..." or "advanced ... methods," you can get a clue as to the number and difficulty of questions to be asked in any given field. More questions, and hence broader coverage, would be included for those subjects which are more important in the work. Now weigh your strengths and weaknesses against the job requirements and prepare accordingly.

3) Determine the level of the position

Another way to tell how intensively you should prepare is to understand the level of the job for which you are applying. Is it the entering level? In other words, is this the position in which beginners in a field of work are hired? Or is it an intermediate or advanced level? Sometimes this is indicated by such words as "Junior" or "Senior" in the class title. Other jurisdictions use Roman numerals to designate the level – Clerk I, Clerk II, for example. The word "Supervisor" sometimes appears in the title. If the level is not indicated by the title, check the description of duties. Will you be working under very close supervision, or will you have responsibility for independent decisions in this work?

4) Choose appropriate study materials

Now that you know the subjects to be examined and the relative amount of each subject to be covered, you can choose suitable study materials. For beginning level jobs, or even advanced ones, if you have a pronounced weakness in some aspect of your training, read a modern, standard textbook in that field. Be sure it is up to date and has general coverage. Such books are normally available at your library, and the librarian will be glad to help you locate one. For entry-level positions, questions of appropriate difficulty are chosen – neither highly advanced questions, nor those too simple. Such questions require careful thought but not advanced training.

If the position for which you are applying is technical or advanced, you will read more advanced, specialized material. If you are already familiar with the basic principles of your field, elementary textbooks would waste your time. Concentrate on advanced textbooks and technical periodicals. Think through the concepts and review difficult problems in your field.

These are all general sources. You can get more ideas on your own initiative, following these leads. For example, training manuals and publications of the government agency which employs workers in your field can be useful, particularly for technical and professional positions. A letter or visit to the government department involved may result in more specific study suggestions, and certainly will provide you with a more definite idea of the exact nature of the position you are seeking.

III. KINDS OF TESTS

Tests are used for purposes other than measuring knowledge and ability to perform specified duties. For some positions, it is equally important to test ability to make adjustments to new situations or to profit from training. In others, basic mental abilities not dependent on information are essential. Questions which test these things may not appear as pertinent to the duties of the position as those which test for knowledge and information. Yet they are often highly important parts of a fair examination. For very general questions, it is almost impossible to help you direct your study efforts. What we can do is to point out some of the more common of these general abilities needed in public service positions and describe some typical questions.

1) General information

Broad, general information has been found useful for predicting job success in some kinds of work. This is tested in a variety of ways, from vocabulary lists to questions about current events. Basic background in some field of work, such as sociology or economics, may be sampled in a group of questions. Often these are principles which have become familiar to most persons through exposure rather than through formal training. It is difficult to advise you how to study for these questions; being alert to the world around you is our best suggestion.

2) Verbal ability

An example of an ability needed in many positions is verbal or language ability. Verbal ability is, in brief, the ability to use and understand words. Vocabulary and grammar tests are typical measures of this ability. Reading comprehension or paragraph interpretation questions are common in many kinds of civil service tests. You are given a paragraph of written material and asked to find its central meaning.

3) Numerical ability

Number skills can be tested by the familiar arithmetic problem, by checking paired lists of numbers to see which are alike and which are different, or by interpreting charts and graphs. In the latter test, a graph may be printed in the test booklet which you are asked to use as the basis for answering questions.

4) Observation

A popular test for law-enforcement positions is the observation test. A picture is shown to you for several minutes, then taken away. Questions about the picture test your ability to observe both details and larger elements.

5) Following directions

In many positions in the public service, the employee must be able to carry out written instructions dependably and accurately. You may be given a chart with several columns, each column listing a variety of information. The questions require you to carry out directions involving the information given in the chart.

6) Skills and aptitudes

Performance tests effectively measure some manual skills and aptitudes. When the skill is one in which you are trained, such as typing or shorthand, you can practice. These tests are often very much like those given in business school or high school courses. For many of the other skills and aptitudes, however, no short-time preparation can be made. Skills and abilities natural to you or that you have developed throughout your lifetime are being tested.

Many of the general questions just described provide all the data needed to answer the questions and ask you to use your reasoning ability to find the answers. Your best preparation for these tests, as well as for tests of facts and ideas, is to be at your physical and mental best. You, no doubt, have your own methods of getting into an exam-taking mood and keeping "in shape." The next section lists some ideas on this subject.

IV. KINDS OF QUESTIONS

Only rarely is the "essay" question, which you answer in narrative form, used in civil service tests. Civil service tests are usually of the short-answer type. Full instructions for answering these questions will be given to you at the examination. But in case this is your first experience with short-answer questions and separate answer sheets, here is what you need to know:

1) Multiple-choice Questions

Most popular of the short-answer questions is the "multiple choice" or "best answer" question. It can be used, for example, to test for factual knowledge, ability to solve problems or judgment in meeting situations found at work.

A multiple-choice question is normally one of three types—

- It can begin with an incomplete statement followed by several possible endings. You are to find the one ending which *best* completes the statement, although some of the others may not be entirely wrong.
- It can also be a complete statement in the form of a question which is answered by choosing one of the statements listed.
- It can be in the form of a problem – again you select the best answer.

Here is an example of a multiple-choice question with a discussion which should give you some clues as to the method for choosing the right answer:

When an employee has a complaint about his assignment, the action which will *best* help him overcome his difficulty is to
- A. discuss his difficulty with his coworkers
- B. take the problem to the head of the organization
- C. take the problem to the person who gave him the assignment
- D. say nothing to anyone about his complaint

In answering this question, you should study each of the choices to find which is best. Consider choice "A" – Certainly an employee may discuss his complaint with fellow employees, but no change or improvement can result, and the complaint remains unresolved. Choice "B" is a poor choice since the head of the organization probably does not know what assignment you have been given, and taking your problem to him is known as "going over the head" of the supervisor. The supervisor, or person who made the assignment, is the person who can clarify it or correct any injustice. Choice "C" is, therefore, correct. To say nothing, as in choice "D," is unwise. Supervisors have and interest in knowing the problems employees are facing, and the employee is seeking a solution to his problem.

2) True/False Questions

The "true/false" or "right/wrong" form of question is sometimes used. Here a complete statement is given. Your job is to decide whether the statement is right or wrong.

SAMPLE: A person-to-person long-distance telephone call costs less than a station-to-station call to the same city.

This statement is wrong, or false, since person-to-person calls are more expensive.

This is not a complete list of all possible question forms, although most of the others are variations of these common types. You will always get complete directions for answering questions. Be sure you understand *how* to mark your answers – ask questions until you do.

V. RECORDING YOUR ANSWERS

For an examination with very few applicants, you may be told to record your answers in the test booklet itself. Separate answer sheets are much more common. If this separate answer sheet is to be scored by machine – and this is often the case – it is highly important that you mark your answers correctly in order to get credit.

An electric scoring machine is often used in civil service offices because of the speed with which papers can be scored. Machine-scored answer sheets must be marked with a pencil, which will be given to you. This pencil has a high graphite content which responds to the electric scoring machine. As a matter of fact, stray dots may register as answers, so do not let your pencil rest on the answer sheet while you are pondering the correct answer. Also, if your pencil lead breaks or is otherwise defective, ask for another.

Since the answer sheet will be dropped in a slot in the scoring machine, be careful not to bend the corners or get the paper crumpled.

The answer sheet normally has five vertical columns of numbers, with 30 numbers to a column. These numbers correspond to the question numbers in your test booklet. After each number, going across the page are four or five pairs of dotted lines. These short dotted lines have small letters or numbers above them. The first two pairs may also have a "T" or "F" above the letters. This indicates that the first two pairs only are to be used if the questions are of the true-false type. If the questions are multiple choice, disregard the "T" and "F" and pay attention only to the small letters or numbers.

Answer your questions in the manner of the sample that follows:

32. The largest city in the United States is
    A. Washington, D.C.
    B. New York City
    C. Chicago
    D. Detroit
    E. San Francisco

1) Choose the answer you think is best. (New York City is the largest, so "B" is correct.)
2) Find the row of dotted lines numbered the same as the question you are answering. (Find row number 32)
3) Find the pair of dotted lines corresponding to the answer. (Find the pair of lines under the mark "B.")
4) Make a solid black mark between the dotted lines.

VI. BEFORE THE TEST

Common sense will help you find procedures to follow to get ready for an examination. Too many of us, however, overlook these sensible measures. Indeed, nervousness and fatigue have been found to be the most serious reasons why applicants fail to do their best on civil service tests. Here is a list of reminders:

- Begin your preparation early – Don't wait until the last minute to go scurrying around for books and materials or to find out what the position is all about.
- Prepare continuously – An hour a night for a week is better than an all-night cram session. This has been definitely established. What is more, a night a

week for a month will return better dividends than crowding your study into a shorter period of time.

- Locate the place of the exam – You have been sent a notice telling you when and where to report for the examination. If the location is in a different town or otherwise unfamiliar to you, it would be well to inquire the best route and learn something about the building.
- Relax the night before the test – Allow your mind to rest. Do not study at all that night. Plan some mild recreation or diversion; then go to bed early and get a good night's sleep.
- Get up early enough to make a leisurely trip to the place for the test – This way unforeseen events, traffic snarls, unfamiliar buildings, etc. will not upset you.
- Dress comfortably – A written test is not a fashion show. You will be known by number and not by name, so wear something comfortable.
- Leave excess paraphernalia at home – Shopping bags and odd bundles will get in your way. You need bring only the items mentioned in the official notice you received; usually everything you need is provided. Do not bring reference books to the exam. They will only confuse those last minutes and be taken away from you when in the test room.
- Arrive somewhat ahead of time – If because of transportation schedules you must get there very early, bring a newspaper or magazine to take your mind off yourself while waiting.
- Locate the examination room – When you have found the proper room, you will be directed to the seat or part of the room where you will sit. Sometimes you are given a sheet of instructions to read while you are waiting. Do not fill out any forms until you are told to do so; just read them and be prepared.
- Relax and prepare to listen to the instructions
- If you have any physical problem that may keep you from doing your best, be sure to tell the test administrator. If you are sick or in poor health, you really cannot do your best on the exam. You can come back and take the test some other time.

VII. AT THE TEST

The day of the test is here and you have the test booklet in your hand. The temptation to get going is very strong. Caution! There is more to success than knowing the right answers. You must know how to identify your papers and understand variations in the type of short-answer question used in this particular examination. Follow these suggestions for maximum results from your efforts:

1) Cooperate with the monitor
The test administrator has a duty to create a situation in which you can be as much at ease as possible. He will give instructions, tell you when to begin, check to see that you are marking your answer sheet correctly, and so on. He is not there to guard you, although he will see that your competitors do not take unfair advantage. He wants to help you do your best.

2) Listen to all instructions
Don't jump the gun! Wait until you understand all directions. In most civil service tests you get more time than you need to answer the questions. So don't be in a hurry.

Read each word of instructions until you clearly understand the meaning. Study the examples, listen to all announcements and follow directions. Ask questions if you do not understand what to do.

3) Identify your papers

Civil service exams are usually identified by number only. You will be assigned a number; you must not put your name on your test papers. Be sure to copy your number correctly. Since more than one exam may be given, copy your exact examination title.

4) Plan your time

Unless you are told that a test is a "speed" or "rate of work" test, speed itself is usually not important. Time enough to answer all the questions will be provided, but this does not mean that you have all day. An overall time limit has been set. Divide the total time (in minutes) by the number of questions to determine the approximate time you have for each question.

5) Do not linger over difficult questions

If you come across a difficult question, mark it with a paper clip (useful to have along) and come back to it when you have been through the booklet. One caution if you do this – be sure to skip a number on your answer sheet as well. Check often to be sure that you have not lost your place and that you are marking in the row numbered the same as the question you are answering.

6) Read the questions

Be sure you know what the question asks! Many capable people are unsuccessful because they failed to *read* the questions correctly.

7) Answer all questions

Unless you have been instructed that a penalty will be deducted for incorrect answers, it is better to guess than to omit a question.

8) Speed tests

It is often better NOT to guess on speed tests. It has been found that on timed tests people are tempted to spend the last few seconds before time is called in marking answers at random – without even reading them – in the hope of picking up a few extra points. To discourage this practice, the instructions may warn you that your score will be "corrected" for guessing. That is, a penalty will be applied. The incorrect answers will be deducted from the correct ones, or some other penalty formula will be used.

9) Review your answers

If you finish before time is called, go back to the questions you guessed or omitted to give them further thought. Review other answers if you have time.

10) Return your test materials

If you are ready to leave before others have finished or time is called, take ALL your materials to the monitor and leave quietly. Never take any test material with you. The monitor can discover whose papers are not complete, and taking a test booklet may be grounds for disqualification.

VIII. EXAMINATION TECHNIQUES

1) Read the general instructions carefully. These are usually printed on the first page of the exam booklet. As a rule, these instructions refer to the timing of the examination; the fact that you should not start work until the signal and must stop work at a signal, etc. If there are any *special* instructions, such as a choice of questions to be answered, make sure that you note this instruction carefully.

2) When you are ready to start work on the examination, that is as soon as the signal has been given, read the instructions to each question booklet, underline any key words or phrases, such as *least, best, outline, describe* and the like. In this way you will tend to answer as requested rather than discover on reviewing your paper that you *listed without describing*, that you selected the *worst* choice rather than the *best* choice, etc.

3) If the examination is of the objective or multiple-choice type – that is, each question will also give a series of possible answers: A, B, C or D, and you are called upon to select the best answer and write the letter next to that answer on your answer paper – it is advisable to start answering each question in turn. There may be anywhere from 50 to 100 such questions in the three or four hours allotted and you can see how much time would be taken if you read through all the questions before beginning to answer any. Furthermore, if you come across a question or group of questions which you know would be difficult to answer, it would undoubtedly affect your handling of all the other questions.

4) If the examination is of the essay type and contains but a few questions, it is a moot point as to whether you should read all the questions before starting to answer any one. Of course, if you are given a choice – say five out of seven and the like – then it is essential to read all the questions so you can eliminate the two that are most difficult. If, however, you are asked to answer all the questions, there may be danger in trying to answer the easiest one first because you may find that you will spend too much time on it. The best technique is to answer the first question, then proceed to the second, etc.

5) Time your answers. Before the exam begins, write down the time it started, then add the time allowed for the examination and write down the time it must be completed, then divide the time available somewhat as follows:
   - If 3-1/2 hours are allowed, that would be 210 minutes. If you have 80 objective-type questions, that would be an average of 2-1/2 minutes per question. Allow yourself no more than 2 minutes per question, or a total of 160 minutes, which will permit about 50 minutes to review.
   - If for the time allotment of 210 minutes there are 7 essay questions to answer, that would average about 30 minutes a question. Give yourself only 25 minutes per question so that you have about 35 minutes to review.

6) The most important instruction is to *read each question* and make sure you know what is wanted. The second most important instruction is to *time yourself properly* so that you answer every question. The third most

9

important instruction is to *answer every question*. Guess if you have to but include something for each question. Remember that you will receive no credit for a blank and will probably receive some credit if you write something in answer to an essay question. If you guess a letter – say "B" for a multiple-choice question – you may have guessed right. If you leave a blank as an answer to a multiple-choice question, the examiners may respect your feelings but it will not add a point to your score. Some exams may penalize you for wrong answers, so in such cases *only*, you may not want to guess unless you have some basis for your answer.

7) Suggestions
   a. Objective-type questions
      1. Examine the question booklet for proper sequence of pages and questions
      2. Read all instructions carefully
      3. Skip any question which seems too difficult; return to it after all other questions have been answered
      4. Apportion your time properly; do not spend too much time on any single question or group of questions
      5. Note and underline key words – *all, most, fewest, least, best, worst, same, opposite*, etc.
      6. Pay particular attention to negatives
      7. Note unusual option, e.g., unduly long, short, complex, different or similar in content to the body of the question
      8. Observe the use of "hedging" words – *probably, may, most likely,* etc.
      9. Make sure that your answer is put next to the same number as the question
      10. Do not second-guess unless you have good reason to believe the second answer is definitely more correct
      11. Cross out original answer if you decide another answer is more accurate; do not erase until you are ready to hand your paper in
      12. Answer all questions; guess unless instructed otherwise
      13. Leave time for review

   b. Essay questions
      1. Read each question carefully
      2. Determine exactly what is wanted. Underline key words or phrases.
      3. Decide on outline or paragraph answer
      4. Include many different points and elements unless asked to develop any one or two points or elements
      5. Show impartiality by giving pros and cons unless directed to select one side only
      6. Make and write down any assumptions you find necessary to answer the questions
      7. Watch your English, grammar, punctuation and choice of words
      8. Time your answers; don't crowd material

8) Answering the essay question

Most essay questions can be answered by framing the specific response around several key words or ideas. Here are a few such key words or ideas:

M's: manpower, materials, methods, money, management
P's: purpose, program, policy, plan, procedure, practice, problems, pitfalls, personnel, public relations

    a. Six basic steps in handling problems:
1. Preliminary plan and background development
2. Collect information, data and facts
3. Analyze and interpret information, data and facts
4. Analyze and develop solutions as well as make recommendations
5. Prepare report and sell recommendations
6. Install recommendations and follow up effectiveness

    b. Pitfalls to avoid
1. *Taking things for granted* – A statement of the situation does not necessarily imply that each of the elements is necessarily true; for example, a complaint may be invalid and biased so that all that can be taken for granted is that a complaint has been registered
2. *Considering only one side of a situation* – Wherever possible, indicate several alternatives and then point out the reasons you selected the best one
3. *Failing to indicate follow up* – Whenever your answer indicates action on your part, make certain that you will take proper follow-up action to see how successful your recommendations, procedures or actions turn out to be
4. *Taking too long in answering any single question* – Remember to time your answers properly

IX. AFTER THE TEST

Scoring procedures differ in detail among civil service jurisdictions although the general principles are the same. Whether the papers are hand-scored or graded by machine we have described, they are nearly always graded by number. That is, the person who marks the paper knows only the number – never the name – of the applicant. Not until all the papers have been graded will they be matched with names. If other tests, such as training and experience or oral interview ratings have been given, scores will be combined. Different parts of the examination usually have different weights. For example, the written test might count 60 percent of the final grade, and a rating of training and experience 40 percent. In many jurisdictions, veterans will have a certain number of points added to their grades.

After the final grade has been determined, the names are placed in grade order and an eligible list is established. There are various methods for resolving ties between those who get the same final grade – probably the most common is to place first the name of the person whose application was received first. Job offers are made from the eligible list in the order the names appear on it. You will be notified of your grade and your rank as soon as all these computations have been made. This will be done as rapidly as possible.

People who are found to meet the requirements in the announcement are called "eligibles." Their names are put on a list of eligible candidates. An eligible's chances of getting a job depend on how high he stands on this list and how fast agencies are filling jobs from the list.

When a job is to be filled from a list of eligibles, the agency asks for the names of people on the list of eligibles for that job. When the civil service commission receives this request, it sends to the agency the names of the three people highest on this list. Or, if the job to be filled has specialized requirements, the office sends the agency the names of the top three persons who meet these requirements from the general list.

The appointing officer makes a choice from among the three people whose names were sent to him. If the selected person accepts the appointment, the names of the others are put back on the list to be considered for future openings.

That is the rule in hiring from all kinds of eligible lists, whether they are for typist, carpenter, chemist, or something else. For every vacancy, the appointing officer has his choice of any one of the top three eligibles on the list. This explains why the person whose name is on top of the list sometimes does not get an appointment when some of the persons lower on the list do. If the appointing officer chooses the second or third eligible, the No. 1 eligible does not get a job at once, but stays on the list until he is appointed or the list is terminated.

X. HOW TO PASS THE INTERVIEW TEST

The examination for which you applied requires an oral interview test. You have already taken the written test and you are now being called for the interview test – the final part of the formal examination.

You may think that it is not possible to prepare for an interview test and that there are no procedures to follow during an interview. Our purpose is to point out some things you can do in advance that will help you and some good rules to follow and pitfalls to avoid while you are being interviewed.

What is an interview supposed to test?

The written examination is designed to test the technical knowledge and competence of the candidate; the oral is designed to evaluate intangible qualities, not readily measured otherwise, and to establish a list showing the relative fitness of each candidate – as measured against his competitors – for the position sought. Scoring is not on the basis of "right" and "wrong," but on a sliding scale of values ranging from "not passable" to "outstanding." As a matter of fact, it is possible to achieve a relatively low score without a single "incorrect" answer because of evident weakness in the qualities being measured.

Occasionally, an examination may consist entirely of an oral test – either an individual or a group oral. In such cases, information is sought concerning the technical knowledges and abilities of the candidate, since there has been no written examination for this purpose. More commonly, however, an oral test is used to supplement a written examination.

Who conducts interviews?

The composition of oral boards varies among different jurisdictions. In nearly all, a representative of the personnel department serves as chairman. One of the members of the board may be a representative of the department in which the candidate would work. In some cases, "outside experts" are used, and, frequently, a businessman or some other representative of the general public is asked to serve. Labor and management or other special groups may be represented. The aim is to secure the services of experts in the appropriate field.

However the board is composed, it is a good idea (and not at all improper or unethical) to ascertain in advance of the interview who the members are and what groups they represent. When you are introduced to them, you will have some idea of their backgrounds and interests, and at least you will not stutter and stammer over their names.

What should be done before the interview?
　　While knowledge about the board members is useful and takes some of the surprise element out of the interview, there is other preparation which is more substantive. It *is* possible to prepare for an oral interview – in several ways:

1) Keep a copy of your application and review it carefully before the interview
　　This may be the only document before the oral board, and the starting point of the interview. Know what education and experience you have listed there, and the sequence and dates of all of it. Sometimes the board will ask you to review the highlights of your experience for them; you should not have to hem and haw doing it.

2) Study the class specification and the examination announcement
　　Usually, the oral board has one or both of these to guide them. The qualities, characteristics or knowledges required by the position sought are stated in these documents. They offer valuable clues as to the nature of the oral interview. For example, if the job involves supervisory responsibilities, the announcement will usually indicate that knowledge of modern supervisory methods and the qualifications of the candidate as a supervisor will be tested. If so, you can expect such questions, frequently in the form of a hypothetical situation which you are expected to solve. NEVER go into an oral without knowledge of the duties and responsibilities of the job you seek.

3) Think through each qualification required
　　Try to visualize the kind of questions you would ask if you were a board member. How well could you answer them? Try especially to appraise your own knowledge and background in each area, *measured against the job sought*, and identify any areas in which you are weak. Be critical and realistic – do not flatter yourself.

4) Do some general reading in areas in which you feel you may be weak
　　For example, if the job involves supervision and your past experience has NOT, some general reading in supervisory methods and practices, particularly in the field of human relations, might be useful. Do NOT study agency procedures or detailed manuals. The oral board will be testing your understanding and capacity, not your memory.

5) Get a good night's sleep and watch your general health and mental attitude
　　You will want a clear head at the interview. Take care of a cold or any other minor ailment, and of course, no hangovers.

What should be done on the day of the interview?
　　Now comes the day of the interview itself. Give yourself plenty of time to get there. Plan to arrive somewhat ahead of the scheduled time, particularly if your appointment is in the fore part of the day. If a previous candidate fails to appear, the board might be ready for you a bit early. By early afternoon an oral board is almost invariably behind schedule if there are many candidates, and you may have to wait.

Take along a book or magazine to read, or your application to review, but leave any extraneous material in the waiting room when you go in for your interview. In any event, relax and compose yourself.

The matter of dress is important. The board is forming impressions about you – from your experience, your manners, your attitude, and your appearance. Give your personal appearance careful attention. Dress your best, but not your flashiest. Choose conservative, appropriate clothing, and be sure it is immaculate. This is a business interview, and your appearance should indicate that you regard it as such. Besides, being well groomed and properly dressed will help boost your confidence.

Sooner or later, someone will call your name and escort you into the interview room. *This is it.* From here on you are on your own. It is too late for any more preparation. But remember, you asked for this opportunity to prove your fitness, and you are here because your request was granted.

What happens when you go in?

The usual sequence of events will be as follows: The clerk (who is often the board stenographer) will introduce you to the chairman of the oral board, who will introduce you to the other members of the board. Acknowledge the introductions before you sit down. Do not be surprised if you find a microphone facing you or a stenotypist sitting by. Oral interviews are usually recorded in the event of an appeal or other review.

Usually the chairman of the board will open the interview by reviewing the highlights of your education and work experience from your application – primarily for the benefit of the other members of the board, as well as to get the material into the record. Do not interrupt or comment unless there is an error or significant misinterpretation; if that is the case, do not hesitate. But do not quibble about insignificant matters. Also, he will usually ask you some question about your education, experience or your present job – partly to get you to start talking and to establish the interviewing "rapport." He may start the actual questioning, or turn it over to one of the other members. Frequently, each member undertakes the questioning on a particular area, one in which he is perhaps most competent, so you can expect each member to participate in the examination. Because time is limited, you may also expect some rather abrupt switches in the direction the questioning takes, so do not be upset by it. Normally, a board member will not pursue a single line of questioning unless he discovers a particular strength or weakness.

After each member has participated, the chairman will usually ask whether any member has any further questions, then will ask you if you have anything you wish to add. Unless you are expecting this question, it may floor you. Worse, it may start you off on an extended, extemporaneous speech. The board is not usually seeking more information. The question is principally to offer you a last opportunity to present further qualifications or to indicate that you have nothing to add. So, if you feel that a significant qualification or characteristic has been overlooked, it is proper to point it out in a sentence or so. Do not compliment the board on the thoroughness of their examination – they have been sketchy, and you know it. If you wish, merely say, "No thank you, I have nothing further to add." This is a point where you can "talk yourself out" of a good impression or fail to present an important bit of information. Remember, *you close the interview yourself.*

The chairman will then say, "That is all, Mr. _____, thank you." Do not be startled; the interview is over, and quicker than you think. Thank him, gather your belongings and take your leave. Save your sigh of relief for the other side of the door.

How to put your best foot forward

Throughout this entire process, you may feel that the board individually and collectively is trying to pierce your defenses, seek out your hidden weaknesses and embarrass and confuse you. Actually, this is not true. They are obliged to make an appraisal of your qualifications for the job you are seeking, and they want to see you in your best light. Remember, they must interview all candidates and a non-cooperative candidate may become a failure in spite of their best efforts to bring out his qualifications. Here are 15 suggestions that will help you:

1) Be natural – Keep your attitude confident, not cocky

If you are not confident that you can do the job, do not expect the board to be. Do not apologize for your weaknesses, try to bring out your strong points. The board is interested in a positive, not negative, presentation. Cockiness will antagonize any board member and make him wonder if you are covering up a weakness by a false show of strength.

2) Get comfortable, but don't lounge or sprawl

Sit erectly but not stiffly. A careless posture may lead the board to conclude that you are careless in other things, or at least that you are not impressed by the importance of the occasion. Either conclusion is natural, even if incorrect. Do not fuss with your clothing, a pencil or an ashtray. Your hands may occasionally be useful to emphasize a point; do not let them become a point of distraction.

3) Do not wisecrack or make small talk

This is a serious situation, and your attitude should show that you consider it as such. Further, the time of the board is limited – they do not want to waste it, and neither should you.

4) Do not exaggerate your experience or abilities

In the first place, from information in the application or other interviews and sources, the board may know more about you than you think. Secondly, you probably will not get away with it. An experienced board is rather adept at spotting such a situation, so do not take the chance.

5) If you know a board member, do not make a point of it, yet do not hide it

Certainly you are not fooling him, and probably not the other members of the board. Do not try to take advantage of your acquaintanceship – it will probably do you little good.

6) Do not dominate the interview

Let the board do that. They will give you the clues – do not assume that you have to do all the talking. Realize that the board has a number of questions to ask you, and do not try to take up all the interview time by showing off your extensive knowledge of the answer to the first one.

7) Be attentive

You only have 20 minutes or so, and you should keep your attention at its sharpest throughout. When a member is addressing a problem or question to you, give him your undivided attention. Address your reply principally to him, but do not exclude the other board members.

8) Do not interrupt

A board member may be stating a problem for you to analyze. He will ask you a question when the time comes. Let him state the problem, and wait for the question.

9) Make sure you understand the question

Do not try to answer until you are sure what the question is. If it is not clear, restate it in your own words or ask the board member to clarify it for you. However, do not haggle about minor elements.

10) Reply promptly but not hastily

A common entry on oral board rating sheets is "candidate responded readily," or "candidate hesitated in replies." Respond as promptly and quickly as you can, but do not jump to a hasty, ill-considered answer.

11) Do not be peremptory in your answers

A brief answer is proper – but do not fire your answer back. That is a losing game from your point of view. The board member can probably ask questions much faster than you can answer them.

12) Do not try to create the answer you think the board member wants

He is interested in what kind of mind you have and how it works – not in playing games. Furthermore, he can usually spot this practice and will actually grade you down on it.

13) Do not switch sides in your reply merely to agree with a board member

Frequently, a member will take a contrary position merely to draw you out and to see if you are willing and able to defend your point of view. Do not start a debate, yet do not surrender a good position. If a position is worth taking, it is worth defending.

14) Do not be afraid to admit an error in judgment if you are shown to be wrong

The board knows that you are forced to reply without any opportunity for careful consideration. Your answer may be demonstrably wrong. If so, admit it and get on with the interview.

15) Do not dwell at length on your present job

The opening question may relate to your present assignment. Answer the question but do not go into an extended discussion. You are being examined for a *new* job, not your present one. As a matter of fact, try to phrase ALL your answers in terms of the job for which you are being examined.

Basis of Rating

Probably you will forget most of these "do's" and "don'ts" when you walk into the oral interview room. Even remembering them all will not ensure you a passing grade. Perhaps you did not have the qualifications in the first place. But remembering them will help you to put your best foot forward, without treading on the toes of the board members.

Rumor and popular opinion to the contrary notwithstanding, an oral board wants you to make the best appearance possible. They know you are under pressure – but they also want to see how you respond to it as a guide to what your reaction would be under the pressures of the job you seek. They will be influenced by the degree of poise you display, the personal traits you show and the manner in which you respond.

EXAMINATION SECTION

EXAMINATION SECTION
TEST 1

DIRECTIONS: Each question or incomplete statement is followed by several suggested answers or completions. Select the one that BEST answers the question or completes the statement. *PRINT THE LETTER OF THE CORRECT ANSWER IN THE SPACE AT THE RIGHT.*

1. The difference between the boiling point and the freezing point of water on the Fahrenheit scale is

    A. 0° B. 100° C. 112° D. 180°

1.____

2. All amino acids contain

    A. calcium and carbon   B. hydrogen and nitrogen
    C. iron and oxygen   D. manganese and phosphorus

2.____

3. Acids and bases combine to form compounds known as

    A. colloids   B. salts   C. solids   D. solutions

3.____

4. $C_6H_{12}O_6$ represents the formula for a(n)

    A. protein   B. salt   C. sugar   D. oil

4.____

5. The soil pH which is suitable for MOST garden crops varies between

    A. 2 and 5   B. 5 and 8   C. 8 and 11   D. 11 and 14

5.____

6. The farmer who plants peas, clover or alfalfa improves the soil PRIMARILY by increasing the available amount of

    A. carbon   B. hydrogen   C. nitrogen   D. oxygen

6.____

7. Phenolphthalein is *generally* used as a(n)

    A. buffer   B. drying agent
    C. emulsifying agent   D. indicator

7.____

8. Of the following, the one classified as a compound is

    A. aluminum   B. ammonia   C. nitrogen   D. sulfur

8.____

9. The process in which a liquid is vaporized and then condensed is called

    A. crystallization   B. decantation
    C. distillation   D. filtration

9.____

10. The formula *4-8-4* used in fertilizers refers to

    A. calcium, magnesium, and sulfur
    B. calcium, nitrogen, and phosphorus
    C. nitrogen, phosphorus, and potassium
    D. nitrogen, potassium, and sodium

10.____

11. The CHIEF source of fuel energy for the living cell are   11._____

    A. carbohydrates and fats
    B. carbohydrates and proteins
    C. fats and proteins
    D. water and carbohydrates

12. The structures of the human alimentary canal, in the order in which food passes through   12._____
    them, are as follows: first the mouth and throat, and then, IN ORDER, the

    A. esophagus, the small intestine, the large intestine, and the stomach
    B. esophagus, the stomach, the large intestine, and the small intestine
    C. esophagus, the stomach, the small intestine, and the large intestine
    D. stomach, the large intestine, the small intestine, and the esophagus

13. Pepsin is a stomach enzyme which   13._____

    A. changes fats to fatty acids
    B. converts starches to sugars
    C. curdles milk
    D. reduces proteins to peptides

14. The substance responsible for the clotting of human blood is known as   14._____

    A. fibrinogen   B. hemoglobin
    C. plasma   D. serum

15. Of the following statements regarding endocrine glands, the one which is NOT true is   15._____
    that

    A. endocrine glands have tubes or ducts to discharge their products to areas of use
    B. hormones are produced in endocrine glands
    C. the adrenal gland is an example of an endocrine gland
    D. the secretions of endocrine glands may be found in the bloodstream

16. The enzyme responsible for breaking fat and fat-like substances into glycerol and fatty   16._____
    acids is

    A. amylase   B. coagulase   C. lipase   D. oxidase

17. An acute x-ray dose of 600 roentgens applied to the entire body is   17._____

    A. insignificant except in the case of persons with an abnormally low level of tolerance
       to x-rays
    B. nearly always fatal
    C. severe in the view of some radiologists and should be avoided as a regular matter
       unless a person is employed as an x-ray technician
    D. tolerable in the average person, but such doses should not be applied more than
       once monthly

18. Radiations may cause cancer, yet radiations are used to treat cancer.   18._____
    This statement is

    A. *false;* radiations cannot cause cancer
    B. *false;* radiations cannot injure cancerous cells

C. *true;* radiations injure malignant cells but not healthy cells
D. *true;* radiations injure malignant cells without doing proportionate harm to non-malignant cells

19. The uranium-238 atom contains 92 protons and 146 neutrons. The number of electrons in the U-238 atom is   19.____

   A. 54   B. 92   C. 146   D. 238

20. The nucleus of the uranium-238 atom contains   20.____

   A. electrons and neutrons   B. electrons and protons
   C. neutrons and protons   D. neutrons only

21. Assume that a sample of radium with an atomic weight of 226 contains 250,000 atoms. Assume further that the half-life ($T_{1/2}$) of the radium is 1,600 years.
   This means MOST NEARLY that in   21.____

   A. 1,600 years 125,000 atoms of the radium sample will have decayed
   B. 1,600 years the portion of the sample which will have decayed can be expressed by the formula $T_{1/2} = 250,000/226$
   C. 3,200 years the portion of the sample which will have decayed can be expressed by the formula $T_{1/2} = 250,000/226$
   D. 3,200 years the radium sample will have decayed completely

22. The scientist who demonstrated that smallpox could be prevented by inoculating the skin of humans with material from cowpox lesions was   22.____

   A. Edward Jenner   B. Robert Koch
   C. Joseph Meister   D. Theodor Schwann

23. Staphylococci appear under microscopic examinations as   23.____

   A. four cells arranged as a square
   B. irregular clusters resembling bunches of grapes
   C. pairs of cells
   D. rows of cells, beadlike or chainlike

24. *Phenol coefficient* refers to a measure of the   24.____

   A. amount of phenol which may be added to food for use as a preservative
   B. effectiveness of a disinfectant in relation to phenol
   C. percentage of carbolic acid found in solutions containing phenol
   D. rapidity with which phenol destroys capsulated bacterial cells

25. The Schick test is used to determine susceptibility to   25.____

   A. diphtheria   B. smallpox
   C. tetanus   D. typhoid fever

26. Infectious hepatitis is a disease caused by   26.____

   A. bacteria   B. protozoa
   C. rickettsiae   D. viruses

27. *Phagocytes* are

    A.   antigens which are used in the production of antibodies
    B.   bacteria which destroy red blood cells
    C.   cells in the human body which protect it from infection
    D.   pathogens which may be present during coughing and sneezing

28. The Breed method is generally used in the bacteriological examination of

    A. meat   B. milk   C. soil   D. water

29. The magnifying power of a microscope may be determined by

    A.   adding the power of the objective to the power of the eyepiece
    B.   dividing the power of the eyepiece into the power of the objective
    C.   multiplying the power of the eyepiece by the power of the objective
    D.   subtracting the power of the eyepiece from the power of the objective

30. The statement regarding viruses which is NOT true is that they

    A.   are all parasites
    B.   are responsible for poliomyelitis
    C.   contain desoxyribonucleic acid
    D.   grow in animals but not in plants

Questions 31-35.

DIRECTIONS: For each of Questions 31 through 35, select the letter preceding the word whose meaning is MOST NEARLY the same as that of the capitalized word.

31. NOXIOUS

    A. gaseous   B. harmful   C. soothing   D. repulsive

32. PYOGENIC

    A. disease producing   B. fever producing
    C. pus forming   D. water forming

33. RENAL

    A. brain   B. heart   C. kidney   D. stomach

34. ENDEMIC

    A. epidemic
    B. endermic
    C. endoblast
    D. peculiar to a particular people or locality, as a disease

35. MACULATION

    A. reticulation   B. inoculation
    C. maturation   D. defilement

KEY (CORRECT ANSWERS)

1.	D		16.	C
2.	B		17.	B
3.	B		18.	D
4.	C		19.	B
5.	B		20.	C
6.	C		21.	A
7.	D		22.	A
8.	B		23.	B
9.	C		24.	B
10.	C		25.	A
11.	A		26.	D
12.	C		27.	C
13.	D		28.	B
14.	A		29.	C
15.	A		30.	D

31.	B
32.	C
33.	C
34.	D
35.	D

TEST 2

DIRECTIONS: Each question or incomplete statement is followed by several suggested answers or completions. Select the one that BEST answers the question or completes the statement. *PRINT THE LETTER OF THE CORRECT ANSWER IN THE SPACE AT THE RIGHT.*

1. The immunity found in individuals who have recovered from measles is termed _____ acquired _____ immunity.

   A. artificially; active
   B. artificially; passive
   C. naturally; active
   D. naturally; passive

   1.____

2. Decomposition of fresh or cold storage meats can be detected BEST by

   A. noting absence of surface moisture
   B. noting presence of *off* odors
   C. noting warmth when touched
   D. observing discoloration

   2.____

3. Bacterial control of shellfish and shellfish growing areas is being based increasingly in this country upon the density of the Escherichia coli organisms in the waters from which shellfish are collected.
   The BEST reason for this is that

   A. E. coli are virulent pathogens which produce serious diseases in man
   B. the density of E. coli in water is relatively easy to determine by shellfish fishermen
   C. the presence of E. coli is an indicator of the presence of human wastes in the water
   D. shellfish which ingest E. coli have objectionable odors which canning cannot remove

   3.____

4. Proper cleaning of dairy utensils entails rinsing with

   A. cold or lukewarm water followed by scrubbing with a detergent solution
   B. cold or lukewarm water followed by scrubbing with hot soapy water
   C. hot water followed by scrubbing with a detergent solution
   D. hot water followed by scrubbing with hot soapy water

   4.____

5. Of the following, the MOST accurate statement regarding the use of chlorine in the purification of public water supplies is:

   A. A small amount of residual chlorine in the water is desirable
   B. Chlorine will destroy most bacteria in the water with the exception of the coliform organisms
   C. The amount of chlorine added to water should be less than the *chlorine demand* of the water
   D. The use of chlorine in public water supplies should be resorted to only in cases of emergency

   5.____

6. Pasteurization entails the heating of milk to AT LEAST _____ for _____ minutes.

   A. 143° F; 15   B. 143° F; 30   C. 161° F; 15   D. 161° F; 30

   6.____

7. The pH value of water is of considerable significance when chlorinating swimming pools.   7.____
   The reason for this is that chlorine functions BEST as a bactericide when the pH value
   of the water is

   A. *high;* also, a high pH water value reduces or prevents eye smarting
   B. *high;* however, a high pH water value increases the possibility of eye smarting
   C. *low;* also, a low pH water value reduces or prevents eye smarting
   D. *low;* however, a low pH water value increases the possibility of eye smarting

8. A test commonly used for determining the presence of chlorine in water is the _____   8.____
   test.

   A. orthotolidine   B. phosphatase
   C. TPI   D. Weil-Felix

9. The chemical which is added to water samples from chlorinated swimming pools to neu-   9.____
   tralize residual chlorine is sodium

   A. bromide   B. carbonate
   C. hydroxide   D. thiosulfate

10. Some years ago, the city experienced an outbreak of food poisoning from potato salad   10.____
    which was kept in an enameled utensil. The vinegar present in the potato salad dis-
    solved a sufficient quantity of a certain substance found in the enamelware to cause
    poisoning. The name of the offending substance was

    A. antimony   B. arsenic   C. cyanide   D. zinc

11. The common housefly, Musca Domestica, is a(n)   11.____

    A. biting insect which does not transmit disease
    B. biting insect which may transmit disease
    C. insect which does not bite and does not transmit disease
    D. insect which does not bite but may transmit disease

12. The name of the substance which it has been suggested be added to *sleeping medicines*   12.____
    to induce vomiting in the event of an overdose is

    A. chlorpromazine   B. ipecac
    C. reserpine   D. seconal

13. The statement which BEST describes DDT is: 13.____
    DDT is

    A. a contact insect poison
    B. an instantaneous poison
    C. effective against all insects
    D. non-toxic to humans

14. The application of 10% DDT dust to rat runways and burrows is   14.____

    A. *advisable,* since it will serve as an effective rodenticide
    B. *advisable,* since it will serve to kill fleas which infest rats
    C. *inadvisable,* since DDT in such amounts stimulates rat growth
    D. *inadvisable,* since rats will be forced to use alternate runways and burrows making
       their elimination more difficult

15. Oligodynamic action refers to the 15.____

    A. ability of extremely small amounts of certain metals to exert a lethal effect upon bacteria
    B. change in levels of chlorine dilution brought about by evaporation
    C. discoloration of tiles in swimming pools due to the excessive mineral content of hard water
    D. removal of organic materials from water by means of sedimentation and filtration

16. The term *BOD,* as used in sewage disposal, refers MOST NEARLY to the 16.____

    A. consumption of oxygen by microorganisms engaged in the decomposition of organic material
    B. contamination of oysters and other shellfish by pathogenic bacteria making them unsafe for human consumption
    C. formation of finely suspended sewage material due to vigorous aeration by powerful pumps
    D. removal of suspended or floating objects from raw sewage by screening

Questions 17-22.

DIRECTIONS: Questions 17 through 22 are based on the Health Code.

17. Assume that the applicant for a Health Department permit is under 21 years of age. The statement which BEST applies to such applicant is: 17.____

    A. Age is not a factor in the issuance of permits
    B. The applicant may be issued a permit provided he is 18 years of age or over if the commissioner waives the age requirement
    C. The establishment of an age requirement for various permits is left solely to the discretion of the commissioner, who may fix any age requirement he deems appropriate
    D. Under no circumstances may a permit be issued to a person under 21 years of age

18. Assume that a person enters a neighborhood pharmacy and asks that a barbiturate be sold to him. He gives the attending pharmacist the name of his physician and states that he does not have the physician's written prescription for such barbiturate with him.
In such a case, the pharmacist 18.____

    A. may dispense a small amount of the barbiturate without requiring a physician's prescription
    B. may dispense the barbiturate in any amount the pharmacist deems reasonable provided the person is either personally known to the pharmacist or presents proper identification
    C. may telephone the physician and accept the physician's oral prescription subject to the physician's later submission of a written prescription
    D. must insist that he be given the physician's written prescription before he dispenses a barbiturate in any quantity

19. A restaurant owner keeps and houses a cat in his restaurant in order to minimize the danger of rat infestation. He also permits patrons to bring their dogs into his restaurant. The CORRECT statement concerning these actions is that the Health Code _____ the owner to keep his cat on the premises _____ visiting the restaurant with their dogs.

    A. permits; and is silent with respect to patrons
    B. permits; but prohibits patrons from
    C. prohibits; and prohibits patrons from
    D. prohibits; but is silent with respect to patrons

19.____

20. The Health Code provides that utensils, such as knives, forks, spoons, cups, and saucers, used in the preparation and service of food are to be cleaned after each use.
The Code provides that such cleaning shall consist of _____ cleaning(s) with a suitable detergent in clean hot water followed by _____ rinsing(s).

    A. *one*; *one*
    B. *one*; *two* successive
    C. *two* successive; *one*
    D. *two* successive; *two* successive

20.____

21. The owner of a meat market uses certain dyes which impart color to meat.
The use of such coloring matter is

    A. absolutely prohibited
    B. permitted if the owner displays a sign which informs consumers that he uses coloring matter
    C. permitted only if the coloring matter is applied to ground beef and to no other meat
    D. prohibited unless such use complies with the provisions of the Federal Meat Inspection Act

21.____

22. Homogenized milk is milk which has been subjected to a treatment so that after 48 hours of quiescent storage the percent of butter fat in the upper one-tenth portion of a container will NOT exceed the percentage of butter fat in the remaining portion of the container by more than

    A. 5%   B. 10%   C. 15%   D. 20%

22.____

23. The Health Code names certain chemicals which, under stated circumstances, may be added to the drinking water supply within a building for anti-corrosion or anti-scaling purposes.
Of the following chemicals, the one which is NOT specifically authorized for this purpose is

    A. calcium bicarbonate   B. calcium hydroxide
    C. sodium carbonate   D. sodium hydroxide

23.____

24. The Code provides that water in swimming pools must meet a certain standard of clarity. This standard is based on the

    A. addition of a chemical to the water which causes a color change if the water does not meet the prescribed standard
    B. measurement by the laboratory of the turbidity of a sample of pool water

24.____

C. use of a black disc, six inches in diameter
D. visual inspection by a sanitarian without the use of any aids or devices

Questions 25-27.

DIRECTIONS: Questions 25 through 27 are to be answered SOLELY on the basis of the following passage.

The first laws prohibiting tampering with foods and selling unwholesome provisions were enacted in ancient times. Early Mosaic and Egyptian laws governed the handling of meat. Greek and Roman laws attempted to prevent the watering of wine. In 200 B.C., India provided for the punishment of adulterators of grains and oils. In the same era, China had agents to prohibit the making of spurious articles and the defrauding of purchasers. Most of our food laws, however, came to us as a heritage from our European forebears.

In early times, foods were few and very simple, and trade existed mostly through barter. Such cheating as did occur was crude and easily detected by the prospective buyer. In the Middle Ages, traders and merchants began to specialize and united themselves into guilds. One of the earliest was called the Pepperers – the spice traders of the day. The Pepperers soon absorbed the grocers and in England got a charter from the king as the Grocer's Company. They set up an ethical code designed to protect the integrity and quality of the spices and other foods sold. Later they appointed a corps of food inspectors to test and certify the merchandise sold to and by the grocers. These men were the first public food inspectors of England.

Pepper is a good example of trade practices that brought about the need for the food inspectors. The demand for pepper was widespread. Its price was high; it was handled by various people during its long journey from the Spice Islands to the grocer's shelf. Each handler had opportunity to debase it; the grinders had the best chance since adulterants could not be detected by methods then available. Worthless barks and seeds, iron ore, charcoal, nutshells, and olive pits were ground along with the berries.

Bread was another food that offered temptation to unscrupulous persons. The most common cheating practice was short weighing but at times the flour used contained ground dried peas or beans.

25. Of the following, the MOST suitable title for the foregoing passage would be: 25.___

   A. Consumer Pressure and Pure Food Laws
   B. Practices Which Brought About the Need for Food Inspectors
   C. Substances Commonly Used as Pepper Adulterants
   D. The Role Played By Pepper as a Spice and as a Preservative

26. The statement BEST supported by the above passage is: 26.___

   A. Food inspectors employed by the Pepperers were responsible for detecting the presence of ground peas in flour
   B. The first guild to be formed in the Middle Ages was known as the Pepperers
   C. The Pepperers were chartered by the king and in accordance with his instructions set up an ethical code
   D. There were persons other than those who handled spices exclusively who became members of the Pepperers

27. The statement BEST supported by the above passage is: 27.___

   A. Early laws of England forbade the addition of adulterants to flour
   B. Egyptian laws of ancient times concerned themselves with meat handling

C. India provided for the punishment of persons adding ground berries and olive pits to spices
D. The Greeks and Romans succeeded in preventing the watering of wine

Questions 28-30.

DIRECTIONS: Questions 28 through 30 are to be answered SOLELY on the basis of the following passage.

Water can purify itself up to a point, by natural processes, but there is a limit to the pollution load that a stream can handle. Self-purification, a complicated process, is brought about by a combination of physical, chemical, and biological factors. The process is the same in all bodies of water, but its intensity is governed by varying environment conditions.

The time required for self-purification is governed by the degree of pollution and the character of the stream. In a large stream, many days of flow may be required for a partial purification. In clean, flowing streams, the water is usually saturated with dissolved purification. In clean, flowing streams, the water is usually saturated with dissolved oxygen, absorbed from the atmosphere and given off by green water plants. The solids of sewage and other wastes are dispersed when they enter the stream and eventually settle. Bacteria in the water and in the wastes themselves begin the process of breaking down the unstable wastes. The process uses up the dissolved oxygen in the water, upon which fish and other aquatic life also depend.

Streams offset the reduction of dissolved oxygen by absorbing it from the air and from oxygen-producing aquatic plants. This replenishment permits the bacteria to continue working on the wastes and the purification process to advance. Replenishment takes place rapidly in a swiftly flowing, turbulent stream because waves provide greater surface areas through which oxygen can be absorbed. Relatively motionless ponds or deep, sluggish streams require more time to renew depleted oxygen.

When large volumes of wastes are discharged into a stream, the water becomes murky. Sunlight no longer penetrates to the water plants, which normally contribute to the oxygen supply through photosynthesis, and the plants die. If the volume of pollution, in relation to the amount of water in the stream and the speed of flow, is so great that the bacteria use the oxygen more rapidly than re-aeration occurs, only putrifying types of bacteria can survive, and the natural process of self-purification is slowed. So the stream becomes foul smelling and looks greasy. Fish and other aquatic life disappear.

28. According to the above passage, if the proportion of wastes to stream water is very high, then the   28.____

    A. amount of dissolved oxygen in the stream increases
    B. death of all bacteria in wastes becomes a certainty
    C. stream will probably look greasy
    D. turbulence of the stream is increased

29. The one of the following which is NOT mentioned in the above passage as a factor in water self-purification is the   29.____

    A. ability of sunlight to penetrate water
    B. percentage of oxygen found in the air
    C. presence of bacteria in waste materials
    D. speed and turbulence of the stream

30. Of the following, the MOST suitable title for the above passage would be: 30.____

    A. Oxygen Requirements of Fish and Other Aquatic Life
    B. Streams as Carriers of Waste Materials
    C. The Function of Bacteria in the Disintegration of Wastes
    D. The Self-purification of Water

Questions 31-32.

DIRECTIONS: Questions 31 and 32 are to be answered SOLELY on the basis of the following passage.

Processing by quick freezing has expanded rapidly. The consumption of frozen fruits and vegetables (on a fresh-equivalent basis) was about 8 pounds per capita annually in the years immediately before the Second World War. It exceeded 200 pounds in 2008.

One example of this growth is frozen concentrated orange juice. From the beginning of commercial production in Florida during the 2005-2006 season, the pack of frozen concentrated orange juice has grown until it amounted to more than 320 million gallons in the 2008-2009 season. That is enough juice, when reconstituted, to supply every person in this country with about 160 average-size servings.

Another striking change in the pattern of food consumption is the sharp increase in consumption of broilers or fryers, young chickens of either sex, usually 8-10 weeks old, and weighing about three pounds.

The commercial production of broilers has increased more than 500 percent since 2006. The number produced exceeded 1.6 billion birds in 2008. On a per capita basis, broiler consumption was about 20 pounds annually (ready-to-cook equivalent basis). This is roughly one-fourth as much as per capita consumption of beef and nearly one-third as large as per capita consumption of pork. Consumption of broilers in the years just after the Second World War was less than one-tenth as large as the consumption of either beef or pork.

Among the factors responsible for this rapid growth are developments in breeding that led to faster gains in weight, lower prices in relation to other meat, and improvements in methods of preparing broilers for market. When broilers, like other poultry, were retailed in an uneviscerated form, dressed broilers could be held for only limited periods. Consequently, birds were shipped to market live, and dressing operations took place mostly in or near terminal markets - the centers of population.

Thus, it is that consumers benefit both from the variety of products available at all seasons of the year and from the many forms in which these products are sold.

31. According to the foregoing passage, the number of broilers produced in 2006 was MOST 31.____
NEARLY

    A. 320,000,000   B. 1,200,000,000
    C. 4,000,000,000   D. 5,200,000,000

32. According to the above passage, the per capita annual consumption of frozen fruits and 32.____
vegetables immediately following the end of World War II

    A. cannot be determined from the above passage
    B. was 16 percent of the per capita consumption of 2008
    C. was most nearly in excess of 200 pounds
    D. was most nearly 8 pounds

Questions 33-35.

DIRECTIONS: For each of Questions 33 through 35, select the letter preceding the word whose meaning is MOST NEARLY the same as that of the capitalized word.

33.   AEROSOL, a _____ dispersed in a _____   33._____

     A.   gas; liquid   B.   liquid; gas
     C.   liquid; solid   D.   solid; liquid

34.   ETIOLOGY   34._____

     A.   cause of a disease   B.   method of cure
     C.   method of diagnosis   D.   study of insects

35.   IN VITRO, in   35._____

     A.   alkali   B.   the body
     C.   the test tube   D.   vacuum

KEY (CORRECT ANSWERS)

1.	C	16.	A
2.	B	17.	B
3.	C	18.	C
4.	A	19.	D
5.	A	20.	B
6.	B	21.	D
7.	D	22.	B
8.	A	23.	A
9.	D	24.	C
10.	A	25.	B
11.	D	26.	D
12.	B	27.	B
13.	A	28.	C
14.	B	29.	B
15.	A	30.	D

31.	B
32.	A
33.	B
34.	A
35.	C

EXAMINATION SECTION
TEST 1

DIRECTIONS: Each question or incomplete statement is followed by several suggested answers or completions. Select the one that BEST answers the question or completes the statement. *PRINT THE LETTER OF THE CORRECT ANSWER IN THE SPACE AT THE RIGHT.*

1. According to Department regulations, whenever meat is packaged by a retailer in advance of being sold, which one of the following MUST also be provided not more than 30 feet from the display counter?
   A. A chart indicating the date the item must be removed from sale
   B. A chart indicating the date the item was first placed on sale
   C. A means of testing the item for adulteration
   D. An accurate computing scale marked "for customer use" or a sign telling customers where such scale is located

1.___

2. According to Department regulations, retail stores are NOT permitted to sell prepackaged meat unless the package is
   A. colorless and transparent
   B. less than one ounce in weight
   C. of a heat-resistant material
   D. open at one end

2.___

3. Hamburger meat may contain all of the following EXCEPT
   A. chemical preservatives   B. added fat
   C. chuck steak   D. neck meat

3.___

4. The net weight declaration on a package of food MUST be
   A. in grams as well as ounces
   B. near the top of the package
   C. on the label but in no specific place
   D. on the main panel of the label

4.___

5. The fat content of oleomargarine MUST be at least
   A. 40 percent   B. 60 percent
   C. 80 percent   D. 90 percent

5.___

6. The following foods contain standardized ingredients EXCEPT
   A. ice cream   B. jams and jellies
   C. ketchup   D. orange drink

6.___

7. Earthenware dishes very often affect food stored in them by being the source of
   A. asbestos contamination B. bacteria
   C. lead contamination   D. fluid dyes

7.___

8. The presence of E. Coli in food PROBABLY means that it   8.___
    A. is contaminated by fecal matter
    B. is high in minerals
    C. is suitable for diabetics
    D. must be refrigerated

9. Botulism food poisoning in the United States is *usually*   9.___
caused by
    A. eating fish caught in polluted waters
    B. failure to wash raw fruit before eating
    C. improper home-canning of fruits and vegetables
    D. tapeworms found in beef or sheep

10. Food poisoning cases in the United States are *usually*   10.___
characterized by
    A. long periods of illness followed by death
    B. long periods of illness rarely followed by death
    C. short periods of illness followed by death
    D. short periods of illness rarely followed by death

11. In the United States, food poisoning due to eating   11.___
mushrooms is LARGELY attributable to
    A. failure to cook mushrooms
    B. failure to wash mushrooms
    C. mushrooms which are blue in color
    D. mushrooms which have not been cultivated domestically

12. Decomposition of fresh or cold storage meats can be   12.___
detected BEST by
    A. noting absence of surface moisture
    B. noting presence of "off" odors
    C. noting warmth when touched
    D. observing discoloration

13. Bacterial control of shellfish and shellfish growing areas   13.___
is being based increasingly in this country upon the
density of the Escherichia coli organisms in the waters
from which shellfish are collected.
The BEST reason for this is that
    A. E. coli are virulent pathogens which produce serious
       diseases in man
    B. the density of E. coli in water is relatively easy to
       determine by shellfish fishermen
    C. the presence of E. coli is an indicator of the presence
       of human wastes in the water
    D. shellfish which ingest E. coli have objectionable
       odors which canning cannot remove

14. Proper cleaning of dairy utensils entails rinsing with   14.___
    A. cold or lukewarm water followed by scrubbing with
       a detergent solution
    B. cold or lukewarm water followed by scrubbing with
       hot soapy water
    C. hot water followed by scrubbing with a detergent
       solution
    D. hot water followed by scrubbing with hot soapy water

15. Of the following foods, the type that is *most likely* to   15.___
    cause "staph" food poisoning if improperly prepared or
    handled is
    A. sugar-coated food   B. dried food
    C. pickled food   D. cream-filled food

16. Harmful bacteria are *most often* introduced into foods   16.___
    prepared in a food service operation by
    A. insects   B. rodents
    C. employees   D. utensils

17. The one of the following procedures that could cause   17.___
    food poisoning is
    A. allowing cooked poultry to stand for an hour, slicing
       it and covering it with broth, and holding it at
       room temperature for several hours
    B. keeping food mixtures on cafeteria counters for one
       hour
    C. cooking left-over food mixtures quickly by frequent
       stirring and then refrigerating in shallow pans
    D. chilling all ingredients for salads at least one
       hour before preparation

18. Trichinosis is a disease which may be caused by   18.___
    A. eating ham which has been overcooked
    B. unsanitary handling of frozen meats
    C. eating food which has been contaminated by infected
       flies
    D. eating infected pork which has been cooked
       insufficiently

19. Of the following, the bacteria which causes MOST food   19.___
    poisoning cases is
    A. botulinum   B. salmonella
    C. pneumococci   D. streptococci

20. Of the following, the BEST reason for discarding the   20.___
    green part of potatoes is that it contains a poison known as
    A. cevitamic acid   B. citric acid
    C. solanine   D. trichinae

21. "Flat sour"   21.___
    A. is spoilage of canned food by bacteria with formation
       of gas
    B. renders food unfit for consumption
    C. can be corrected by addition of sugar to food before
       serving
    D. should be re-boiled before serving

22. Trichinosis is a disease caused by   22.___
    A. a worm   B. an allergy
    C. improper refrigeration   D. food adulteration

23. The ONLY safe method of canning non-acid vegetables and   23.___
    meats is the
    A. open kettle   B. hot water bath
    C. pressure process   D. cold pack

24. Spoilage in canned foods which is caused by bacteria that   24.____
produces acid without gas is known as
    A. putrefaction   B. fermentation
    C. botulinus   D. flat-sour spoilage

25. To avoid the development of bacterial toxins in custards   25.____
and cream pies, one should
    A. cool to room temperature before refrigeration
    B. refrigerate within half hour after cooking
    C. heat to 212° F. during cooking
    D. store in the freezing compartment

26. An excellent medium for the growth of bacteria which cause   26.____
food poisoning toxins is
    A. cream puffs   B. pickled watermelon rind
    C. nougat candies   D. preserves

27. Cooked foods should be cooled and refrigerated quickly   27.____
PRIMARILY to
    A. prevent growth and development of bacteria
    B. preserve food nutrients
    C. prevent loss of moisture content
    D. preserve a fresh-cooked appearance

28. Aerobic bacteria which cause food spoilage   28.____
    A. are unable to grow without air
    B. are able to grow without air
    C. grow equally well with or without air
    D. need heat and moisture for growth

29. A disease caused by contamination in canned foods is   29.____
    A. trichinosis   B. botulism
    C. undulant fever   D. tularemia

30. Oysters which feed on sewage sometimes transmit   30.____
    A. rabies   B. yellow fever
    C. typhoid fever   D. malaria

31. In order to retard spoilage of bread, many baking   31.____
companies add
    A. sodium sulphathionate
    B. sodium propionate
    C. sodium hypophosphate
    D. sodium benzoate

32. Spoilage in canned foods may be caused by   32.____
    A. filling the jars with food and fluid even with the top
    B. allowing the jars to cool before sealing the jars
       completely
    C. heating the jars for use in the hot-pack method
    D. filling the jars with the liquid in which the food
       was cooked

33. To prevent curdling of mayonnaise,   33.____
    A. expose to light   B. expose to air
    C. store at 32° F.   D. store at 150° F.

34. When a retailer plans to offer for sale thawed meat or   34.___
    fish, he is required by Department regulations to do which
    one of the following?
    A. Label the product "thawed" or "defrosted"
    B. Reduce the price of the product
    C. Refreeze the product and label it "refrozen"
    D. Remove the unsold portion from sale within three hours

35. Certain perishable foods must be stamped, printed, or other- 35.___
    wise plainly and conspicuously marked with either the last
    day or date of sale or the last day or date of recommended
    usage. Among these foods are
    A. bread, meat and poultry
    B. bread, milk and meat
    C. eggs, bread and milk
    D. eggs, milk and poultry

———

KEY (CORRECT ANSWERS)

1. D	11. D	21. B	31. B
2. A	12. B	22. A	32. B
3. A	13. C	23. C	33. A
4. D	14. A	24. D	34. C
5. C	15. D	25. B	35. C
6. D	16. C	26. A	
7. C	17. A	27. A	
8. A	18. D	28. A	
9. C	19. B	29. B	
10. D	20. C	30. C	

———

TEST 2

DIRECTIONS: Each question or incomplete statement is followed by several suggested answers or completions. Select the one that BEST answers the question or completes the statement. *PRINT THE LETTER OF THE CORRECT ANSWER IN THE SPACE AT THE RIGHT.*

1. Trichinae are destroyed by   1.___
   A. freezing and storing at 15°F
   B. curing in a 2.5% salt solution
   C. radiation sterilization
   D. heating to 125°F

2. Dishes used by a patient with a communicable disease   2.___
   should be
   A. *boiled* for 5 minutes in soapy water
   B. *boiled* in an antiseptic solution
   C. *washed* for 5 minutes in soapy hot water
   D. *washed* in clear water at 180°F

3. The medium of infection which is MOST difficult to control 3.___
   is
   A. insects   B. food   C. water   D. air

4. Bread spoilage is retarded by the addition of   4.___
   A. sodium carbonate   B. calcium propionate
   C. tartaric acid   D. protease

5. Frozen foods which have partially thawed   5.___
   A. may be refrozen
   B. may be cooked and refrozen
   C. must be discarded
   D. may be refrozen only after complete thawing

6. Pasteurization of milk   6.___
   A. kills pathogenic bacteria
   B. retards the growth of bacteria
   C. kills all bacteria
   D. homogenizes

7. Among the following food additives, the one which is used 7.___
   for the purpose of enhancing the keeping quality of the
   food is
   A. vitamin D in milk
   B. bleaching agents in flour
   C. scorbic acid in cider
   D. minerals and vitamins in cereals

8. An example of the bactericidal method of food preservation   8.___
   is
   A. jam and jellies   B. pickling
   C. freezing   D. refrigeration

9. Oysters which feed on sewage sometimes transmit   9.___
   A. rabies   B. yellow fever
   C. typhoid fever   D. malaria

10. The ONLY edible mussel that is sold is the   10.___
    A. scampi   B. scallop
    C. clam   D. rock lobster

11. *Flat sour*   11.___
    A. is spoilage of canned food by bacteria with formation
       of gas
    B. renders food unfit for consumption
    C. can be corrected by addition of sugar to food before
       serving
    D. should be re-boiled before serving

12. Trichinosis is a disease caused by   12.___
    A. a worm   B. an allergy
    C. improper refrigeration D. food adulteration

13. Dry foods should be stored in   13.___
    A. a cool dry place
    B. the basement
    C. a cabinet near the stove
    D. the refrigerator

14. In the process of food preservation,   14.___
    A. all bacteria are destroyed
    B. harmful bacteria are destroyed
    C. the growth of bacteria may be prevented or checked
    D. harmless bacteria are destroyed

15. Orange juice prepared the night before it is to be   15.___
    served should be stored
    A. in a container that will protect it from exposure
       to air
    B. at 32°F
    C. at 70°F
    D. in a plastic shaker-type container

16. When food has been spilled on an electric cooking   16.___
    element,
    A. clean immediately
    B. wash with soap and water when cool
    C. clean with steel wool
    D. clean with a dry brush after food chars

17. Pork should always be cooked to the well-done state in order to
    A. develop the best possible flavor
    B. prevent trichinosis in the consumer
    C. improve the tenderness
    D. prevent loss of nutritives in juices

17.___

18. To prevent curdling of mayonnaise,
    A. expose to light   B. expose to air
    C. store at 32°F   D. store at 150°F

18.___

19. To avoid the development of bacterial toxins in custards and cream pies, one should
    A. cool to room temperature before refrigeration
    B. refrigerate within half hour after cooking
    C. heat to 212°F during cooking
    D. store in the freezing compartment

19.___

20. An excellent medium for the growth of bacteria which cause food poisoning toxins is
    A. cream puffs   B. pickled watermelon rind
    C. nougat candies   D. preserves

20.___

21. The flavor of fruit is due to
    A. its color pigmentation   B. citric and malic acids
    C. inorganic salts   D. pectins

21.___

22. Which of the following is used to ripen fruits and vegetables?
    A. Chlorophyll   B. Methylene
    C. Ethylene   D. Benzoate of soda

22.___

23. Of the following, the BEST selection of orange for making orange juice is the
    A. Rome Beauty   B. Navel
    C. Valencia   D. MacIntosh

23.___

24. To preserve the shape, fruits should be cooked
    A. without sugar
    B. with very little sugar
    C. by adding sugar after cooking
    D. by adding sugar before cooking

24.___

25. A prolific source of pectin for use in industry is
    A. fruits   B. carrots
    C. walnuts   D. calves' knuckles

25.___

26. Substances in fruits and vegetables which are responsible for the ripening process are
    A. molds   B. yeasts   C. bacteria   D. enzymes

26.___

27. Sulfuring dried fruits
    A. promotes retention of vitamin B
    B. prevents darkening
    C. activates vitamin C
    D. increases tenderness

27.___

28. Little spoilage occurs in stored, sun-dried fruits because the
    A. micro-organisms have been destroyed
    B. moisture content is low
    C. pectin is inactive
    D. yeasts do not flourish in the absence of light

28.___

29. Tenderized dried fruits have been
    A. sulphurized, dried, then partially cooked
    B. dried, partially cooked, then partially dried
    C. partially cooked, dried, then partially cooked
    D. dried, sulphurized, then partially cooked

29.___

30. Salted fish roe is sold as
    A. macedoine B. curry   C. brioche   D. caviar

30.___

31. Aerobic bacteria which cause food spoilage
    A. are unable to grow without air
    B. are able to grow without air
    C. grow equally well with or without air
    D. need heat and moisture for growth

31.___

32. For everyday use, the Fahrenheit temperature of the refrigerator should be
    A. 20-25°   B. 35-40°   C. 45-50°   D. 55-60°

32.___

33. Incompletely cooked pork, if eaten, may result in
    A. botulism   B. ptomaine
    C. trichinosis   D. typhoid

33.___

34. The process which makes it possible to store fresh food in any climate without refrigeration for an unlimited length of time is
    A. dehydration   B. freezing
    C. freeze-drying   D. flake-drying

34.___

35. Frozen foods deteriorate in flavor unless they are kept at
    A. 32°F   B. 32°C   C. 0°F   D. 0°C

35.___

KEY (CORRECT ANSWERS)

1. C	11. B	21. D	31. A
2. A	12. A	22. C	32. B
3. D	13. A	23. C	33. C
4. B	14. C	24. D	34. C
5. B	15. A	25. A	35. C
6. A	16. D	26. D	
7. C	17. B	27. B	
8. A	18. A	28. B	
9. C	19. B	29. B	
10. B	20. A	30. D	

EXAMINATION SECTION
TEST 1

DIRECTIONS: Each question or incomplete statement is followed by several suggested answers or completions. Select the one that BEST answers the question or completes the statement. *PRINT THE LETTER OF THE CORRECT ANSWER IN THE SPACE AT THE RIGHT.*

1. Assume that you have been assigned to inspect a building reported to be infested by rats and to prepare a written report thereon.
    Of the following items covered in the report, the LEAST important one is *probably* the   1.____

    A. fact that rats appear to be feeding on the garbage of a luncheonette which adjoins the building
    B. name and address of the building owner
    C. record of past violations by the owner
    D. statement made by tenants regarding the presence of rats

2. After completing an inspection of a food manufacturing plant, you submit a report of your findings to your supervisor. A few days later, you receive a memorandum from your supervisor indicating that the head of the bureau found your report inadequate. You are to re-inspect the establishment immediately. Your supervisor's memorandum lists the areas which he feels your report did not cover adequately. You, however, are convinced that your report is adequate.
    The BEST course of action for you to take at this time is to   2.____

    A. refrain from re-inspecting the food establishment unless directed to do so personally by the head of the bureau
    B. re-inspect the premises, submit another report, and then discuss the matter with your supervisor
    C. telephone your supervisor and insist that the matter be fully discussed before you proceed further with a re-inspection
    D. write a letter to the head of the bureau explaining why you feel your report was adequate, and wait for a reply before you re-inspect

3. Assume that you have a close relative who is engaged in the practice of accounting. Following your inspection of a restaurant which is not in violation of the health code, you inform the owner that your relative is an accountant. You hand the owner the accountant's business card and suggest that your relative be considered for any accounting work needed. The owner then tells you that he would like to have your relative take over his accounting work.
    Your action in securing the restaurant's accounting work for your relative is   3.____

    A. *improper;* you should have discussed the matter with the restaurant owner after your regular working hours
    B. *improper;* you should not have suggested your relative for the owner's accounting work
    C. *proper* as long as the owner remains in full compliance with the health code
    D. *proper* provided that your relative does not discuss the owner's business with you

4. A tenant of an apartment house telephones the department of health to complain that no heat is being furnished to her apartment. The complaint is referred to you with instructions to make a field visit. When you arrive at the apartment house, the tenant partly opens her door but refuses to allow you to enter the apartment. You explain the situation to the tenant, but she persists in her refusal to allow you to enter the apartment.
The BEST thing for you to do in these circumstances is to

    4._____

    A. notify the tenant that if she refuses you admittance to her apartment, you may be required to obtain a court order directing her to allow you to enter
    B. place the complaint in your pending file and return to the apartment the next time you are in the neighborhood
    C. prepare a report setting forth that the tenant refused to allow you to enter the apartment
    D. take a reading of the temperature in the hallway and then estimate the temperature in the apartment

5. In the course of your inspection of a luncheonette, you note a violation of a provision of the health code relating to the unsanitary condition of food containers. You point out the condition to the owner as you begin to prepare a notice of violation. The owner becomes very angry and declares that the food containers are clean. To illustrate his point, he shows the food containers to two patrons seated at the lunch counter. Both patrons declare that the food containers are clean and suggest that you not *pick* on the owner. The owner then tells you that if you make trouble for him, he will make trouble for you.
Of the following, the BEST course of action for you to take is to

    5._____

    A. inform the owner that you will return at a later date to complete your notice of violation
    B. refrain from giving the owner a notice of violation since he has witnesses to support his position
    C. serve the owner with a notice of violation
    D. telephone your supervisor, tell him of the condition of the food containers, and ask him whether you should give the owner a notice of violation

6. A provision of the health code requires food handlers to take a course in food handling sanitation. Your supervisor requests that when you visit food establishments in your district, you remind them of the code requirement. Your supervisor stresses that your visit is to be an educational one and that you are not to emphasize the mandatory aspect of this provision. Later, you visit a restaurant owner in your district who expresses strong reservations as to the practicability of releasing food handlers to take such a course.
The one statement which you should NOT make to the owner under any circumstances is that if his food handlers take such a course,

    6._____

    A. future violations of the health code by the owner will receive special treatment since he is cooperating with the department
    B. his profits may rise since patrons prefer to eat in a place where food sanitation standards are high
    C. the possibility of food poisoning with attendant possible economic loss to the owner will be decreased
    D. the requirement of the health code is mandatory in this respect and must be complied with

7. During your inspection of a multiple dwelling, you find a serious violation of a provision of 7._____
the health code. The owner claims that at one time the particular provision in question
was sensible, but circumstances have changed and the provision should now be
repealed. After listening to the owner, you are convinced that the health code should be
changed as indicated by the owner. The CORRECT course of action for you to take is to

   A. give the owner a notice of violation and refrain from making any report to your
   office concerning the provision in question
   B. give the owner a notice of violation and suggest to your superior that the provision
   be reviewed as to its continued usefulness
   C. refrain from giving the owner a notice of violation since the provision is obviously
   outdated
   D. refrain from giving the owner a notice of violation until the courts rule on the consti-
   tutionality of the provision

8. Assume that you are in the apartment of a tenant who has complained that the landlord 8._____
is not furnishing sufficient heat. Your thermometer shows that the landlord is furnishing
sufficient heat to comply with the pertinent provision of the health code. You so inform the
tenant. The tenant excitedly declares that you are using a *fake* thermometer and that you
may be on the landlord's *payroll*.
Under these circumstances, you should state that

   A. if the tenant has any allegation to make concerning your inspection or character,
   she should contact your department
   B. if these, allegations are repeated, you will refer the tenant for psychiatric examination
   C. the allegations constitute defamation of the character of a public officer, and that
   you will so notify the police department
   D. you will ask the landlord to speak to the tenant to vouch for your honesty

9. You have been assigned to investigate a complaint with regard to a certain fruit and veg- 9._____
etable stand. Your investigation does not disclose any violation. Upon informing the
owner of the stand of your findings, he offers you a bag of fruit as a gift. You decline it. He
then offers to sell you the bag of fruit below the retail price - at cost to him. You SHOULD

   A. accept the offer, but refrain from visiting the establishment again
   B. accept the offer, provided you are satisfied that the fruit is being sold to you at cost
   C. decline the offer because it is not possible to calculate the wholesale cost of the
   fruit
   D. decline the offer since acceptance would be improper

10. The term *FT/SEC* is a unit of 10._____

   A. density   B. length   C. mass   D. speed

11. A container can hold 100 pounds of water or 70 pounds of an *unknown* liquid. 11._____
The specific gravity of the *unknown* liquid is

   A. .30   B. .70   C. 1.0   D. 1.4

12. A *calorie* may be defined as the amount of heat required to raise one 12._____

   A. gram of water 1° C   B. gram of water 1° F
   C. pound of water 1° C   D. pound of water 1° F

13. The acidity of vinegar is due to the presence of acid.   13.____

    A. acetic   B. carbonic   C. citric   D. hydrochloric

14. The cleansing action of a soap solution is due PRIMARILY to its   14.____

    A. acid reaction   B. increased surface tension
    C. neutral reaction   D. reduced surface tension

15. Titration refers to a process of   15.____

    A. determining the normality of an acid solution
    B. determining the refractive index of a crystal
    C. extracting oxygen from water
    D. measuring the quantity of salt present in a saline solution

16. Which one of the following types of compounds ALWAYS includes carbon, hydrogen, and   16.____
oxygen?

    A. Carbohydrates   B. Carbonates
    C. Hydrates   D. Hydrocarbons

17. The formula for nitric acid is   17.____

    A. HNO_2 B. HNO_3 C. NO_2 D. N_2O

18. Gastric juice owes its acidity, *for the most part,* to the presence of _____ acid. 18.____

    A. carbonic   B. hydrochloric   C. nitric   D. sulfuric

19. Insulin is a type of   19.____

    A. enzyme   B. hormone   C. sugar   D. vitamin

20. The organ which prevents food from entering the windpipe during the act of swallowing is   20.____
the

    A. epiglottis   B. larynx   C. pharynx   D. trachea

21. Casein is a type of   21.____

    A. carbohydrate   B. enzyme   C. fat   D. protein

22. The MAIN function of the kidneys is to remove wastes formed as a result of the oxidation   22.____
of

    A. carbohydrates   B. fats   C. proteins   D. vitamins

23. Vitamin C is ALSO known as _____ acid.   23.____

    A. ascorbic   B. citric   C. glutamic   D. lactic

24. Light passes through the crystalline lens in the eye and focuses on the   24.____

    A. cornea   B. iris   C. pupil   D. retina

25. An electron weighs   25.____

    A. less than a neutron   B. more than a neutron
    C. the same as a neutron   D. the same as a proton

KEY (CORRECT ANSWERS)

1.	C		11.	B
2.	B		12.	A
3.	B		13.	A
4.	C		14.	D
5.	C		15.	A
6.	A		16.	A
7.	B		17.	B
8.	A		18.	B
9.	D		19.	B
10.	D		20.	A

21.	D
22.	C
23.	A
24.	D
25.	A

TEST 2

DIRECTIONS: Each question or incomplete statement is followed by several suggested answers or completions. Select the one that BEST answers the question or completes the statement. *PRINT THE LETTER OF THE CORRECT ANSWER IN THE SPACE AT THE RIGHT.*

1. An electron has a _____ charge.   1._____

    A. negative   B. positive   C. variable   D. zero

2. Isotopes are atoms of elements which have _____ atomic weight(s).   2._____

    A. different atomic numbers and different
    B. different atomic numbers but the same
    C. the same atomic number and the same
    D. the same atomic number but different

3. In the Einstein equation $E = mc^2$, E, m, and c^2 stand for, respectively,   3._____

    A. electrons, molecules, and (centimeters)^2
    B. energy, mass, and (light velocity)^2
    C. energy, mass, and (radioactivity)^2
    D. energy, molecules, and (light velocity)^2

4. Photosynthesis entails the absorption of   4._____

    A. carbon dioxide and oxygen and release of water
    B. carbon dioxide and water and release of oxygen
    C. oxygen and release of carbon dioxide and water
    D. water and release of carbon dioxide and oxygen

5. Ordinary body temperature is approximately 37 on the _____ scale.   5._____

    A. absolute   B. A.P.I.   C. centigrade   D. Fahrenheit

6. Bacteria are _____ chlorphyll.   6._____

    A. multicellular organisms containing
    B. multicellular organisms that do not contain
    C. unicellular organisms containing
    D. unicellular organisms that do not contain

7. The immunity acquired as a result of an injection of tetanus antitoxin is termed _____   7._____
immunity.

    A. artificially acquired active
    B. artificially acquired passive
    C. naturally acquired active
    D. naturally acquired passive

8. A virus is the causative agent of   8._____

    A. diphtheria   B. smallpox   C. syphilis   D. tuberculosis

9. Typhus fever epidemics are caused by

    A. bacteria   B. rickettsiae   C. viruses   D. yeasts

9.____

10. The one of the following tests used to determine susceptibility to scarlet fever is the _____ test.

    A. Dick   B. Schick   C. Wasserman   D. Widal

10.____

11. Generally, the type of individual immunity to disease which is of the LONGEST duration is brought about by

    A. antibody production stimulated by killed microorganisms
    B. antibody production stimulated by live microorganisms
    C. transfer of antibodies during pregnancy from an immune mother to her unborn child by placental transfer
    D. transfer of antibodies from one adult to another

11.____

12. Diabetes is considered to be a(n) _____ disease.

    A. communicable   B. contagious
    C. noninfectious   D. infectious

12.____

13. The genus *Mycobacterium* contains a species responsible for

    A. diphtheria   B. gonorrhea
    C. tuberculosis   D. whooping cough

13.____

14. The pH of a neutral solution is

    A. 3   B. 5   C. 7   D. 9

14.____

15. Of the following, the pair that is NOT a set of equivalents is

    A. .014% .00014   B. 1/5% .002
    C. 1.5% 3/200   D. 115% .115

15.____

16. 10^{-2} is equal to

    A. 0.001   B. 0.01   C. 0.1   D. 100.0

16.____

17. $10^2 \times 10^3$ is equal to

    A. 10^5 B. 10^6 C. 100^5 D. 100^6

17.____

18. The length of two objects are in the ratio of 2:1. If each were 3 inches shorter, the ratio would be 3:1. The longer object is _____ inches.

    A. 8   B. 10   C. 12   D. 14

18.____

19. If the weight of water is 62.4 pounds per cubic foot, the weight of the water that fills a rectangular container 6 inches by 6 inches by 1 foot is _____ pounds.

    A. 7.8   B. 15.6   C. 31.2   D. 46.8

19.____

20. *Dry-ice* is solid

    A. ammonia   B. carbon dioxide
    C. freon   D. sulfur dioxide

20.____

21. The fat content of normal milk is *approximately*   21.____

    A. 1%   B. 4%   C. 10%   D. 16%

22. The one of the following acids GENERALLY responsible for the natural souring of milk is   22.____
_____ acid.

    A. acetic   B. amino   C. citric   D. lactic

23. From a nutritional standpoint, milk is *deficient* in   23.____

    A. iron   B. lactose
    C. mineral salts   D. protein

24. The man who is USUALLY known as the father of chemotherapy is   24.____

    A. Paul Ehrlich   B. Elie Metchnikoff
    C. Louis Pasteur   D. John Tyndall

25. The success of this country in building the Panama Canal was due to the successful con-   25.____
quest of yellow fever.
The man who directed the study which led to this conquest was

    A. Joseph Lister   B. Walter Reed
    C. Theobold Smith   D. William Welch

––––––

KEY (CORRECT ANSWERS)

1.	A		11.	B
2.	D		12.	C
3.	B		13.	C
4.	B		14.	C
5.	C		15.	D
6.	D		16.	B
7.	B		17.	A
8.	B		18.	C
9.	B		19.	B
10.	A		20.	B

21. B
22. D
23. A
24. A
25. B

––––––

TEST 3

DIRECTIONS: Each question or incomplete statement is followed by several suggested answers or completions. Select the one that BEST answers the question or completes the statement. *PRINT THE LETTER OF THE CORRECT ANSWER IN THE SPACE AT THE RIGHT.*

1. The *Babcock test* is used in milk analysis to determine _____ content.   1._____

   A. butterfat   B. mineral   C. protein   D. vitamin

2. The phosphatase test is used to determine whether milk   2._____

   A. has an objectionable odor
   B. has been adequately pasteurized
   C. has been adulterated
   D. is too alkaline

3. A lactometer is used in milk inspection work to determine the   3._____

   A. acidity of milk   B. color of milk
   C. percentage of milk solids   D. specific gravity of milk

4. Milk samples collected at milk plants should be taken from milk cans, the contents of which have   4._____

   A. not been stirred so that sediment does not appear in the sample
   B. not been stirred so that the growth of bacteria which thrive on oxygen is not encouraged
   C. been stirred in order to obtain a representative sample
   D. been stirred so that the percentage of dissolved oxygen meets required standards

5. In the holding process, milk should be pasteurized for at least 30 minutes at a temperature of about   5._____

   A. 115° F   B. 145° F   C. 180° F   D. 212° F

6. Undulant fever, which may be contracted from milk, is caused by an organism known as   6._____

   A. Bacillus subtilis   B. Brucella abortus
   C. Staphylococcus aureus   D. Streptococcus pyogenes

7. The presence of *milk stone* or *water stone* in dairy equipment is   7._____

   A. *desirable;* it indicates that dairy equipment is modern
   B. *desirable;* it indicates that milking machines have been sterilized
   C. *undesirable;* it will increase the bacterial count of milk that comes in contact with it
   D. *undesirable;* it will greatly increase the percentage of water in the final milk product

8. The type of dairy barn flooring which is LEAST desirable from a sanitarian's point of view is   8._____

   A. asphalt   B. compressed cork and asphalt
   C. concrete   D. wood

9. *Curds* and *whey* are substances encountered in the manufacture of cheese. Of the two substances, usually one   9._____

A. is made into cheese; the other is a by-product used to feed animals
B. is made into cheese; the other is made into butter
C. is made into hard cheese; the other is made into soft cheese
D. refers to bacteria-ripened cheese; the other refers to mold-ripened cheese

10. Botulism food poisoning in the United States is USUALLY caused by   10.____

   A. eating fish caught in polluted waters
   B. failure to wash raw fruit before eating
   C. improper home-canning of fruits and vegetables
   D. tapeworms found in beef or sheep

11. The growth of pathogenic bacteria in preserved dates and figs is *inhibited* because these   11.____
   foods have a high _____ content.

   A. acid   B. mineral   C. protein   D. sugar

12. In the heating of the following foods during canning, the one which generally requires the   12.____
   LOWEST temperature to prevent microbiological activity is

   A. fish   B. fruit   C. meat   D. milk

13. Food poisoning cases in the United States are USUALLY characterized by _____ fol-   13.____
   lowed by death.

   A. long periods of illness
   B. long periods of illness rarely
   C. short periods of illness
   D. short periods of illness rarely

14. In the United States, food poisoning due to eating mushrooms is LARGELY attributable   14.____
   to

   A. failure to cook mushrooms
   B. failure to wash mushrooms
   C. mushrooms which are blue in color
   D. mushrooms which have not been cultivated domestically

15. Of the following, the food whose flavor is NOT improved by the addition of monosodium   15.____
   glutamate is

   A. cooked vegetables   B. fruit juice
   C. meats   D. seafood and chowders

16. A NEW method of food preservation involves preservation by   16.____

   A. chemicals   B. drying   C. heat   D. radiation

17. In grading meat, the term *finish* refers to   17.____

   A. distribution of fat   B. muscle hardness
   C. presence of tapeworm   D. symmetry of the carcass

18. Of the following preservatives, the one which may NOT be legally used in the preservation of meat is

    A. benzoic acid   B. salt
    C. sugar   D. wood smoke

18.____

19. A vitamin known to be effective in the prevention of pellagra is

    A. ascorbic acid   B. niacin
    C. riboflavin   D. thiamin

19.____

20. Eggs are *candled* for the purpose of determining

    A. calcium content
    B. size of the egg
    C. the presence of blood spots
    D. weight of the egg

20.____

KEY (CORRECT ANSWERS)

1.	A	11.	D
2.	B	12.	B
3.	D	13.	D
4.	C	14.	D
5.	B	15.	B
6.	B	16.	D
7.	C	17.	A
8.	D	18.	A
9.	A	19.	B
10.	C	20.	C

TEST 4

DIRECTIONS: Each question or incomplete statement is followed by several suggested answers or completions. Select the one that BEST answers the question or completes the statement. *PRINT THE LETTER OF THE CORRECT ANSWER IN THE SPACE AT THE RIGHT.*

1. Foodstuffs such as cereal and flour do not readily spoil as a result of bacterial action because such foodstuffs usually have a low _____ content.

    A. acid   B. ash   C. sodium   D. water

1._____

2. The presence of bacteria responsible for typhoid fever in a public water supply is PROBABLY traceable to

    A. fecal contamination
    B. excessive water aeration
    C. pus from skin lesions
    D. rotting animal and fish remains

2._____

3. Objectionable tastes and odors in public water supplies are, in the great majority of cases, due to the presence of

    A. algae and protozoa   B. animal remains
    C. dissolved oxygen   D. yeasts and molds

3._____

4. Atmospheric pressure as indicated by the mercury barometer at sea level is GENERALLY about _____ inches.

    A. 10   B. 15   C. 30   D. 45

4._____

5. The CHIEF objective of a sewage treatment and disposal system is to

    A. alter sewage by chemical treatment so that it may be sold as commercial fertilizer
    B. convert liquid sludge so that it may be used as drinking water
    C. convert sewage into a form usable as land fill
    D. remove or decompose the organic matter

5._____

6. *Warfarin* is GENERALLY used in the control of

    A. ants   B. flies   C. lice   D. rats

6._____

7. The control of the common housefly has been regarded as important because houseflies

    A. are a great nuisance although they are not responsible for the transmission of diseases
    B. may transmit diseases by biting humans
    C. may transmit diseases by contaminating food with pathogenic organisms
    D. may transmit diseases by injecting pathogenic organisms into the bloodstream of animals which are later eaten by man

7._____

8. The term *Anopheles* refers to a type of

    A. ant   B. louse   C. mosquito   D. termite

8._____

9. Galvanized iron is made by coating iron with　　　　　　　　　　　　　　　9.____

    A.   chromium   B.   lead   C.   tin   D.   zinc

10. The amount of oxygen in the air of a properly ventilated room, expressed as a percent-　　10.____
age of volume, is APPROXIMATELY

    A.   5%   B.   10%   C.   15%   D.   20%

11. Field control of hay fever generally depends upon the effective use of a(n)　　　　　11.____

    A.   bacteriostatic agent   B.   fungicide
    C.   insect spray   D.   weed killer

12. An orthotolidine testing set may be used to determine the presence of　　　　　　12.____

    A.   bacterial growth in milk cans and pails
    B.   chlorine in wash and rinse waters
    C.   DDT dust in foods such as flour and sugar
    D.   organisms responsible for the spoilage of shucked oysters

13. The one of the following which is NOT a characteristic of carbon monoxide gas is that it　　13.____

    A.   causes nausea and vomiting
    B.   has a strong irritating odor
    C.   interferes with the oxygen-carrying power of the blood
    D.   is a common constituent of manufactured gas

Question 14.

DIRECTIONS: Question 14 is based on the following statement.

The rise of science is the most important fact of modern life. No student should be per-
mitted to complete his education without understanding it. From a scientific education, we
may expect an understanding of science. From scientific investigation, we may expect scien-
tific knowledge. We are confusing the issue and demanding what we have no right to ask if
we seek to learn from science the goals of human life and of organized society.

14. The foregoing statement implies MOST NEARLY that　　　　　　　　　　　14.____

    A.   in a democratic society, the student must determine whether to pursue a scientific
        education
    B.   organized society must learn from science how to meet the needs of modern life
    C.   science is of great value in molding the character and values of the student
    D.   scientific education is likely to lead the student to acquire an understanding of sci-
        entific processes

Questions 15-16.

DIRECTIONS: Questions 15 and 16 are based on the following statement.

Since sewage is a variable mixture of substances from many sources, it is to be
expected that its microbial flora will fluctuate both in types and numbers. Raw sewage may
contain millions of bacteria per milliliter. Prominent among these are the coliforms. strepto-

cocci, anaerobic spore forming bacilli, the Proteus group, and other types which have their origin in the intestinal tract of man. Sewage is also a potential source of pathogenic intestinal organisms. The poliomyelitis virus has been demonstrated to occur in sewage; other viruses are readily isolated from the same source. Aside from the examination of sewage to demonstrate the presence of some specific microorganism for epidemiological purposes, bacteriological analysis provides little useful information because of the magnitude of variations known to occur with regard to both numbers and kinds.

15. According to the above passage,   15.____

    A. all sewage contains pathogenic organisms
    B. bacteriological analysis of sewage is routinely performed in order to determine the presence of coliform organisms
    C. microorganisms found in sewage vary from time to time
    D. poliomyelitis epidemics are due to viruses found in sewage

16. The title which would be MOST suitable for the above passage is: 16.____

    A. Disposal of Sewage by Bacteria
    B. Microbes and Sewage Treatment
    C. Microbiological Characteristics of Sewage
    D. Sewage Removal Processes

Questions 17-18.

DIRECTIONS: Questions 17 and 18 are based on the following statement.

Most cities carrying on public health work exercise varying degrees of inspection and control over their milk supplies. In some cases, it consists only of ordinances, with little or no attempt at enforcement. In other cases, good control is obtained through wise ordinances and an efficient inspecting force and laboratory. While inspection alone can do much toward controlling the quality and production of milk, there must also be frequent laboratory tests of the milk.

The bacterial count of the milk indicates the condition of the dairy and the methods of milk handling. The counts, therefore, are a check on the reports of the sanitarian. High bacterial counts of milk from a dairy reported by a sanitarian to be "good" may indicate difficulty not suspected by the sanitarian such as infected udders, inefficient sterilisation of utensils, or poor cooling.

17. According to the above passage, the MOST accurate of the following statements is: 17.____

    A. The bacterial count of milk will be low if milk-producing animals are free from disease.
    B. A high bacterial count of milk can be reduced by pasteurization.
    C. The bacterial count of milk can be controlled by the laboratory.
    D. The bacterial count of milk will be low if the conditions of milk production, processing and handling are good.

18. The following conclusion may be drawn from the above passage: 18._____

    A. Large centers of urban population usually exercise complete control over their milk supplies.

    B. Adequate legislation is an important adjunct of a milk supply control program.

    C. Most cities should request the assistance of other cities prior to instituting a milk supply control program.

    D. Wise laws establishing a milk supply control program obviate the need for the enforcement of such laws provided that good laboratory techniques are employed.

Question 19-20.

DIRECTIONS: Questions 19 and 20 are based upon the following excerpt from the health code.

Article 101 Shellfish and Fish
Section 101. 03 Shippers of shellfish; registration
    (a) No shellfish shall be shipped into the city unless the shipper of such shellfish is regis-tered with the department.
    (b) Application for registration shall be made on a form furnished by the department.
    (c) The following shippers shall be eligible to apply for registration :
        1. A shipper of shellfish located in the state but outside the city who holds a shellfish shipper's permit issued by the state conservation department; or
        2. A shipper of shellfish located outside the state, or located in Canada, who holds a shellfish certificate of approval or a permit issued by the state or provincial agency having control of the shellfish industry of his state or province, which certificate of approval or permit appears on the current list of interstate shellfish shipper permits published by the United States Public Health Service.
    (d) The commissioner may refuse to accept the registration of any applicant whose past observance of the shellfish regulations is not satisfactory to the commissioner.
    (e) No applicant shall ship shellfish into the city unless he has been notified in writing by the department that his application for registration has been approved.
    (f) Every registration as a shipper of shellfish, unless sooner revoked, shall terminate on the expiration date of the registrant's state shellfish certificate or permit.

19. The above excerpt from the health code provides that   19._____

    A. permission to register may not be denied to a shellfish shipper meeting the stan-dards of his own jurisdiction

    B. permission to register will not be denied unless the shipper's past observances of shellfish regulations has not been satisfactory to the U.S. Public Health Service

    C. the commissioner may suspend the regulations applicable to registration if requested to do so by the governmental agency having jurisdiction over the shell-fish shipper

    D. an applicant for registration as a shellfish shipper may ship shellfish into the city when notified by the department in writing that his application has been approved

20. The above excerpt from the health code provides that

    A. applications for registration will not be granted to out-of-state shippers of shellfish who have already received permission to sell shellfish from another jurisdiction

    B. shippers of shellfish located outside of the city may not ship shellfish into the city unless the shellfish have passed inspection by the jurisdiction in which the shellfish shipper is located

    C. a shipper of shellfish located in Canada is eligible for registration provided that he holds a shellfish permit issued by the appropriate provincial agency and that such permit appears on the current list of shellfish shipper permits published by the U.S. Public Health Service

    D. a shipper of shellfish located in Canada whose shellfish permit has been revoked by the provincial agency may ship shellfish into the city until such time as he is notified in writing by the department that his shellfish registration has been revoked

20.____

KEY (CORRECT ANSWERS)

1.	D	11.	D
2.	A	12.	B
3.	A	13.	B
4.	C	14.	D
5.	D	15.	C
6.	D	16.	C
7.	C	17.	D
8.	C	18.	B
9.	D	19.	D
10.	D	20.	C

EXAMINATION SECTION
TEST 1

DIRECTIONS: Each question or incomplete statement is followed by several suggested answers or completions. Select the one that BEST answers the question or completes the statement. *PRINT THE LETTER OF THE CORRECT ANSWER IN THE SPACE AT THE RIGHT.*

Questions 1-17.

DIRECTIONS: Questions 1 through 17 pertain to the bacteriological examinations of water.

1. Glassware used in the bacteriological examination of water is GENERALLY sterilized in the oven for at least _____ hour at _____ °C.

    A. $\frac{1}{2}$; 150
    B. $\frac{1}{2}$; 170
    C. 1; 150
    D. 1; 170

1.____

2. Nutrient broth used in the bacteriological examination of water is GENERALLY sterilized at _____ °C for _____ minutes.

    A. 100; 20   B. 100; 90   C. 121; 15   D. 121; 60

2.____

3. Water samples for bacteriological examination should be stored at a temperature between

    A. 0-10° C   B. 10-20° C   C. 20-30° C   D. 30-40° C

3.____

4. Of the following types of glass bottles, the one PREFERRED for use as sample bottles in the bacteriological examination of water is _____ mouth, _____ stoppered bottles.

    A. narrow; ground glass
    B. narrow; rubber
    C. wide; ground glass
    D. wide; rubber

4.____

5. The one of the following NOT recommended for use in culture media to be used in the bacteriological examination of water is

    A. agar
    B. beef extract
    C. meat infusion
    D. peptone

5.____

6. The standard tests used to determine the presence of members of the coliform group in water employ an incubating temperature of

    A. 20° C   B. 20° F   C. 35° C   D. 35° F

6.____

7. Water samples containing residual chlorine are GENERALLY dechlorined by the addition of

    A. phosphoric acid
    B. potassium iodide
    C. sodium phosphate
    D. sodium thiosulfate

7.____

8. The pH of sterilized nutrient broth used in the bacteriological examination of water is APPROXIMATELY

    A. 6.0   B. 6.8   C. 7.4   D. 8.2

8.____

9. The magnification of the lens GENERALLY used in counting colonies for the standard plate count is *approximately* _____ diameters.

    A. 1.5   B. 3.5   C. 5.0   D. 10.0

9.____

10. The medium used for the standard plate count is

    A. brilliant green lactose bile broth
    B. E.M.B. agar
    C. Endo medium
    D. tryptone glucose extract agar

10.____

11. The one of the following tests which is NOT used to differentiate members of the coliform group is the _____ test.

    A. indol   B. methyl red
    C. methylene blue reduction   D. sodium citrate

11.____

12. The one of the following tests which may be used to differentiate members of the coliform group is the _____ test.

    A. Alsterberg   B. Eijkman
    C. Eraser   D. Kjeldahl

12.____

13. A presumptive test is NEGATIVE if gas is not produced within _____ hours.

    A. 12   B. 24   C. 48   D. 72

13.____

14. Agar slants are used in the

    A. completed test   B. confirmed test
    C. presumptive test   D. standard plate count

14.____

15. Primary fermentation tubes are used in the

    A. completed test   B. confirmed test
    C. presumptive test   D. standard plate count

15.____

16. The determination of the Gram-stain characteristics of the organisms isolated is a phase of the

    A. completed test   B. confirmed test
    C. presumptive test   D. standard plate count

16.____

17. The use of formate ricinoleate broth is a phase of the

    A. completed test   B. confirmed test
    C. presumptive test   D. standard plate count

17.____

18. A water supply containing pathogenic microorganisms is termed

    A. contaminated   B. hard
    C. polluted   D. potable

18.____

19. In water analysis, the number and kind of bacteria in the water are determined by _____ examination.

    A. bacteriological   B. chemical
    C. microscopical   D. physical

19.____

20. Of the following methods of water treatment, the one which is used to insure the destruc-   20.____
tion of all pathogenic microorganisms is

    A. chlorination   B. coagulation
    C. filtration   D. sedimentation

21. The British practice for the full bacteriological examination of water includes a test for   21.____

    A. Cl. welchii   B. L. acidophilus
    C. K. pyogenes   D. S. marcescens

22. The CHIEF object of the microscopic examination of water is the determination of the   22.____
presence of microorganisms which produce

    A. *carbonate* and *non-carbonate* hardness
    B. diseases
    C. objectionable turbidity
    D. objectionable tastes and odors

23. Of the following compounds, the one which causes temporary hardness in water is cal-   23.____
cium

    A. bicarbonate   B. chloride
    C. sulfate   D. sulfide

24. The one of the following dissolved gaseous impurities of water which is NOT corrosive to   24.____
metals is

    A. carbon dioxide   B. hydrogen sulfide
    C. nitrogen   D. oxygen

25. A substance COMMONLY found in water which causes mottled enamel of teeth is   25.____
sodium

    A. bicarbonate   B. chloride
    C. fluoride   D. sulfate

KEY (CORRECT ANSWERS)

1.	D	11.	C
2.	C	12.	D
3.	A	13.	C
4.	C	14.	A
5.	C	15.	C
6.	C	16.	A
7.	D	17.	A
8.	B	18.	A
9.	A	19.	A
10.	D	20.	A

21.	A
22.	D
23.	A
24.	C
25.	C

———

READING COMPREHENSION
UNDERSTANDING AND INTERPRETING WRITTEN MATERIAL
EXAMINATION SECTION
TEST 1

DIRECTIONS: Each question or incomplete statement is followed by several suggested answers or completions. Select the one that BEST answers the question or completes the statement. *PRINT THE LETTER OF THE CORRECT ANSWER IN THE SPACE AT THE RIGHT.*

Questions 1-2.

DIRECTIONS: Questions 1 and 2 are to be answered SOLELY on the basis of the following paragraphs.

Another striking change in the pattern of food consumption is the sharp increase in consumption of broilers or fryers, young chickens of either sex, usually 8-10 weeks old, and weighing about three pounds.

The commercial production of broilers increased more than 500 percent from 1960 to 1970. The number produced exceeded 1.6 billion birds in 1970. On a per capita basis, broiler consumption was about 20 pounds annually (ready-to-cook equivalent basis). This is roughly one-fourth as much as per capital consumption of beef and nearly one-third as large as per capita consumption of pork. Consumption of broilers in the years just after the Second World War was less than one-tenth as large as the consumption of either beef or pork.

Among the factors responsible for this rapid growth were developments in breeding that led to faster gains in weight, lower prices in relation to other meat, and improvements in methods of preparing broilers for market. When broilers, like other poultry, were retailed in an uneviscerated form, dressed broilers could be held for only limited periods. Consequently, birds were shipped to market live, and dressing operations took place mostly in or near terminal markets — the centers of population.

Thus, it is that consumers benefit both from the variety of products available at all seasons of the year and from the many forms in which these products are sold.

1. According to the foregoing passage, the number of broilers produced in 1960 was MOST NEARLY   1.____

    A. 80,000,000   B. 300,000,000
    C. 1,000,000,000   D. 1,300,000,000

2. According to the above passage, the per capita annual consumption of frozen fruits and vegetables immediately following the end of World War II   2.____

    A. cannot be determined from the above passage
    B. was 4 percent of the per capita consumption of 1970
    C. was most nearly in excess of 50 pounds
    D. was most nearly two pounds

Questions 3-5.

DIRECTIONS: Questions 3 through 5 are to be answered SOLELY on the basis of the following passage.

The first laws prohibiting tampering with foods and selling unwholesome provisions were
enacted in ancient times. Early Mosaic and Egyptian laws governed the handling of meat.
Greek and Roman laws attempted to prevent the watering of wine. In 200 B.C., India provided
for the punishment of adulterators of grains and oils. In the same era, China had agents to
prohibit the making of spurious articles and the defrauding of purchasers. Most of our food
laws, however, came to us as a heritage from our European forebears.

In early times foods were few and very simple, and trade existed mostly through barter.
Such cheating as did occur was crude and easily detected by the prospective buyer. In the
Middle Ages, traders and merchants began to specialize and united themselves into guilds.
One of the earliest was called the Pepperers—the spice traders of the day. The Pepperers
soon absorbed the grocers and in England got a charter from the king as the Grocer's Company. They set up an ethical code designed to protect the integrity and quality of the spices
and other foods sold. Later they appointed a corps of food inspectors to test and certify the
merchandise sold to and by the grocers. These men were the first public food inspectors of
England.

Pepper is a good example of trade practices that brought about the need for the food
inspectors. The demand for pepper was widespread. Its price was high; it was handled by
various people during its long journey from the Spice Islands to the grocer's shelf. Each handler had opportunity to debase it; the grinders had the best chance since adulterants could
not be detected by methods then available. Worthless barks and seeds, iron ore, charcoal,
nutshells, and olive pits were ground along with the berries.

Bread was another food that offered temptation to unscrupulous persons. The most common cheating practice was short-weighting, but at times the flour used contained ground
dried peas or beans.

3.   Of the following, the MOST suitable title for the foregoing passage would be: 3.___

    A.   CONSUMER PRESSURE AND PURE FOOD LAWS
    B.   PRACTICES WHICH BROUGHT ABOUT THE NEED FOR FOOD INSPECTORS
    C.   SUBSTANCES COMMONLY USED AS PEPPER ADULTERANTS
    D.   THE ROLE PLAYED BY PEPPER AS A SPICE AND AS A PRESERVATIVE

4.   The statement BEST supported by the above passage is: 4.___

    A.   Food inspectors employed by the Pepperers were responsible for detecting the
presence of ground peas in flour.
    B.   The first guild to be formed in the Middle Ages was known as the Pepperers.
    C.   The Pepperers were chartered by the king and in accordance with his instructions
set up an ethical code.
    D.   There were persons other than those who handled spices exclusively who became
members of the Pepperers.

5. The statement BEST supported by the above passage is: 5.____

    A. Early laws of England forbade the addition of adulterants to flour.
    B. Egyptian laws of ancient times concerned themselves with meat handling.
    C. India provided for the punishment of persons adding ground berries and olive pits to spices.
    D. The Greeks and Romans succeeded in preventing the watering of wine.

Questions 6-8.

DIRECTIONS: Questions 6 through 8 are to be answered SOLELY on the basis of the following passage.

Water can purify itself up to a point, by natural processes, but there is a limit to the pollution load that a stream can handle. Self-purification, a complicated process, is brought about by a combination of physical, chemical, and biological factors. The process is the same in all bodies of water, but its intensity is governed by varying environment conditions.

The time required for self-purification is governed by the degree of pollution and the character of the stream. In a large stream, many days of flow may be required for a partial purification. In clean, flowing streams, the water is usually saturated with dissolved purification. In clean, flowing streams, the water is usually saturated with dissolved oxygen, absorbed from the atmosphere and given off by green water plants. The solids of sewage and other wastes are dispersed when they enter the stream and eventually settle. Bacteria in the water and in the wastes themselves begin the process of breaking down the unstable wastes. The process uses up the dissolved oxygen in the water upon which fish and other aquatic life also depend.

Streams offset the reduction of dissolved oxygen by absorbing it from the air and from oxygen-producing aquatic plants. This replenishment permits the bacteria to continue working on the wastes and the purification process to advance. Replenishment takes place rapidly in a swiftly flowing, turbulent stream because waves provide greater surface areas through which oxygen can be absorbed. Relatively motionless ponds or deep, sluggish streams require more time to renew depleted oxygen.

When large volumes of wastes are discharged into a stream, the water becomes murky. Sunlight no longer penetrates to the water plants, which normally contribute to the oxygen supply through photosynthesis, and the plants die. If the volume of pollution, in relation to the amount of water in the stream and the speed of flow, is so great that the bacteria use the oxygen more rapidly than re-aeration occurs, only purifying types of bacteria can survive, and the natural process of self-purification is slowed. So the stream becomes foul smelling and looks greasy. Fish and other aquatic life disappear.

6. According to the above passage, if the proportion of wastes to stream water is very high, then the 6.____

    A. amount of dissolved oxygen in the stream increases
    B. death of all bacteria in wastes becomes a certainty
    C. stream will probably look greasy
    D. turbulence of the stream is increased

7. The one of the following which is NOT mentioned in the above passage as a factor in water self-purification is the

    A. ability of sunlight to penetrate water
    B. percentage of oxygen found in the air
    C. presence of bacteria in waste materials
    D. speed and turbulence of the stream

7.____

8. Of the following, the MOST suitable title for the above passage would be:

    A. OXYGEN REQUIREMENTS OF FISH AND OTHER AQUATIC LIFE
    B. STREAMS AS CARRIERS OF WASTE MATERIALS
    C. THE FUNCTION OF BACTERIA IN THE DISINTEGRATION OF WASTES
    D. THE SELF-PURIFICATION OF WATER

8.____

Question 9.

DIRECTIONS: Question 9 is to be answered SOLELY on the basis of the following passage from the Health Code.

A drug or device shall be deemed to be misbranded:

1. if any word, statement, or other information required by this article to appear on the label or labeling is not prominently placed thereon with such conspicuousness, as compared with other words, statements, designs, or emblems in the labeling, and in such terms as to render it likely to be read and understood by the ordinary individual under customary conditions of purchase and use; or

2. if it is a drug and is not designated solely by a name recognized in an official compendium unless its label bears the common or usual name of the drug, if it has one, and, if it is fabricated from two or more ingredients, the common or usual name of each active ingredient, including the kind and quantity by percentage or amount of any alcohol; or

3. unless its labeling bears adequate directions for use, except that a drug or device may be exempted from this requirement by the Commissioner when he finds that it is not necessary for the protection of the public health, and such adequate warnings against use in those pathological conditions or by children where its use may be dangerous to health, or against unsafe dosage or methods or duration of administration or application, in such manner and form, as are necessary for the protection of users.

9. According to the above passage, the LEAST accurate of the following statements is:

    A. Certain drugs must have labels which give their names as found in an official compendium.
    B. Drugs or devices are not necessarily misbranded if their labels carry warnings against use in certain pathological conditions.
    C. Labels on drugs liable to deterioration must state the precautions necessary to prevent deterioration.
    D. Required information on a drug label should be at least as conspicuous as other statements on the label.

9.____

Questions 10-11.

DIRECTIONS: Questions 10 through 11 are to be answered SOLELY on the basis of the following passage from the Health Code.

(a)   The Commissioner shall not consent to the use or proposed use of a food additive
if the data before him:

(1)   fail to establish that the proposed use of the additive under the conditions of
use specified will be safe; or
(2)   show that the proposed use of the additive would promote deception of the
consumer in violation of this Code, or would otherwise result in adulteration
or in misbranding of food within the meaning of this Code.

(b)   If, in the opinion of the Commissioner, based on the data before him, a tolerance
limitation is required in order to assure that the proposed use of an additive will be safe, the
Commissioner:

(1)   shall not fix such tolerance limitation at a level higher than he finds to be reasonably required to accomplish the physical or other technical effect for
which such additive is intended; and
(2)   shall not consent to the proposed use of the additive if he finds that the data
do not establish that such use would accomplish the intended physical or
other technical effect

(c)   In determining whether a proposed use of a food additive is safe, the Commissioner shall consider among other relevant factors:

(1)   the probable consumption of the additive and of any sub- stance formed in
the food because of the use of the additive; and
(2)   the cumulative effect of such additive in the diet of man or animals.

10.   If the data indicate that the proposed use of a food additive will be safe if the amount   10._____
added is limited to 5 milligrams per gram of the food, the commissioner shall fix the tolerance limitation at

A.   5 milligrams per gram of the food
B.   4 milligrams per gram of the food if this is the amount that can be expected to pro-
duce the desired effect
C.   less than 5 but more than 4 milligrams per gram of the food if 4 milligrams is the
amount that can be expected to produce the desired effect
D.   less than 5 milligrams per gram of the food

11.   According to the above passage, the LEAST accurate of the following statements is: 11._____

A.   Some food additives may, in some cases, be considered as adulterants.
B.   The Commissioner should consider all relevant factors in determining whether the
proposed use of a food additive is safe.

C. The Commissioner may not prohibit the use of an additive if the data show that its use is safe within certain tolerance limitations.

D. The Commissioner may prohibit the use of an additive even if the data indicate that its use would be safe within certain tolerance limitations.

12. Of the following, a LIKELY reason for the inclusion of section (b)(2) given above is that   12._____

    A. food additives used within their tolerance limitations are likely to be unsafe

    B. producers may tend to add more than the safe amount if the tolerance limitation does not permit the accomplishment of the intended physical or other technical effect

    C. the probable consumption of the additive cannot be determined if it does not accomplish the intended physical or other technical effect

    D. use of a food additive that does not accomplish the intended physical or other technical effect is uneconomical

Questions 13-14.

DIRECTIONS: Questions 13 and 14 are to be answered SOLELY on the basis of the following passage from the Health Code.

The new Health Code governs such aspects of the food industry as pertain to cleanliness of apparatus, equipment, and utensils used in the preparation and service of food and sanitation of food establishment premises.

This revision marks a considerable shift in emphasis from detailed specific standards to broad performance standards and the imposition of greater obligation on industry to carry out well-formulated inspection procedures under its own direction and under continuing supervision of the Department. The emphasis is on clean and sanitary food products produced, sold, or served in clean and sanitary food establishments.

The emphasis on generalized performance standards serves the important purpose of encouraging, through less restrictive regulations, the development of new processes in food sanitation and food manufacture. The advances in food technology practices, new chemical aids, and new sanitary designs of machinery have already pointed the way to getting the job done better and without the need for restrictive detailed regulations. This article is not only designed to permit progress of this kind to the fullest, but it also reflects the view that such industry growth should receive constant stimulation so that there is Less need for official policing and more and more self-sanitation and self-supervision.

13. According to the above passage, the new Health Code   13._____

    A. requires detailed specific standards rather than broad performance standards

    B. is intended to provide for ultimate complete self-supervision by the food industry

    C. places less emphasis on self-inspection than on generalized performance standards

    D. is designed to take cognizance of the effects of new developments on food industry practices

14. According to the above passage, the new Health Code does NOT   14._____

    A. consider continued supervision of the food industry by the Department to be of great importance

    B. consider that advances in food technology indicate the need for less restrictive regulations

    C. emphasize coercion but seeks voluntary compliance by the food industry

    D. obligate the food industry to carry out well-formulated inspection procedures under its own direction

Questions 15-16.

DIRECTIONS: Questions 15 and 16 are to be answered SOLELY on the basis of the following passage.

    The beginnings of hygiene can be traced back to antiquity in the sanitary laws of the Hebrews. Preventive medicine began with the first primitive idea of contagion. Even in the time when epidemics were explained as due to the wrath of the gods or visitations of evil spirits, it was observed that certain illnesses apparently spread from person to person. Gradually, the idea of contagiousness was associated with a number of diseases. Fracastorium, in his book, DE CONTAGIONE, published in 1554, proposed a classification of diseases into those which were contagious and those which were not. For three centuries following this publication, the medical profession was divided into two camps: the non-contagionists, who believed that the causative agents of epidemic disease were inanimate and gaseous in nature and associated with emanations from decomposing organic matter, effluvia, and miasma; and the much smaller group, the contagionists, who identified contagiousness with germs of some kind.

    Looking backward, this confusion is understandable. That some diseases were contagious was fairly obvious, but some apparently arose spontaneously without a traceable source. The confusion was finally resolved in the latter part of the nineteenth century by the work of Pasteur, Koch, and their followers. The causative relationship of specific microorganisms for one after another of the infectious diseases was established and the part played by carriers, missed cases, common water and food supplies, arthropod vectors, and animal reservoirs in transmission was gradually elucidated.

15. The above passage IMPLIES that   15._____

    A. all infectious diseases were highly contagious

    B. the contagionists of the early 19th century had identified the specific microorganisms causing certain diseases

    C. the role of animal reservoirs contributed to the confusion which once existed concerning disease transmission

    D. the sanitary laws of the ancient Hebrews show that they had some scientific knowledge of the causes of disease

16. According to the above passage, the MOST accurate of the following statements is: 16._____

    A. Fracastorius believed that all diseases could be caused by miasma.

    B. It is still believed by scientists that certain infectious diseases arise spontaneously.

C. Nothing was accomplished in disease prevention until the germ theory was established.
D. Preventive medicine was practiced to some extent in early times even though epidemics may have been attributed to evil spirits.

Questions 17-18.

DIRECTIONS: Questions 17 and 18 are to be answered SOLELY on the basis of the following paragraph.

Microorganisms are living things so small that they can be seen only with the aid of a microscope. They are widely distributed in nature and are responsible for many physical and chemical changes of importance to the life of plants, of animals, and of human beings. Altogether too many people believe that all *microbes* or *germs* are harmful, and that they are an entirely undesirable group of living things. While it is true that some microorganisms produce disease, the great majority of them do not. In fact, the activities of these hosts of non-disease-producing microorganisms make possible the continued existence of plants and animals on the earth. In addition, many kinds of microorganisms are used in industries to manufacture products of great value to man. But the activities of non-disease-producing microorganisms are not always desirable. Fabrics and fibers may be rotted, fermentation processes may be upset by undesirable organisms, and other harmful effects may occur. From a practical point of view, we are interested in the microorganisms because of the things that they do and the physical and chemical changes which they produce. Also, we are interested in ways and means to control undesirable organisms and to put the useful ones to work; but a study of the activities and the means for control of microorganisms must be based upon knowledge of their nature and life processes.

17. The one of the following which is the MOST suitable title for the above paragraph is   17.____

    A.   BACTERIA CAN BE USEFUL
    B.   MICROORGANISMS AND THE PUBLIC HEALTH SANITARIAN
    C.   THE CONTROL OF MICROBES
    D.   THE RELATIONSHIP OF MICROORGANISMS TO MAN AND HIS ENVIRONMENT

18. According to the above paragraph, the MOST accurate of the following statements is: 18.____

    A.   All non-disease-producing microorganisms are beneficial to mankind.
    B.   *Microbes* or *germs* are terms which are synonymous with *bacteria*.
    C.   The activities of useful bacteria need no controls.
    D.   Without microorganisms, life on earth would be virtually impossible.

Question 19.

DIRECTIONS: Question 19 is to be answered SOLELY on the basis of the following paragraph.

The rise of science is the most important fact of modern life. No student should be permitted to complete his education without understanding it. From a scientific education, we may expect an understanding of science. From scientific investigation, we may expect scientific knowledge. We are confusing the issue and demanding what we have no right to ask if we seek to learn from science the goals of human life and of organized society.

19. The foregoing paragraph implies MOST NEARLY that

    19.____

    A. in a democratic society the student must determine whether to pursue a scientific education
    B. organized society must learn from science how to meet the needs of modern life
    C. science is of great value in molding the character and values of the student
    D. scientific education is likely to lead the student to acquire an understanding of scientific processes

Questions 20-21.

DIRECTIONS: Questions 20 and 21 are to be answered SOLELY on the basis of the following paragraph.

Since sewage is a variable mixture of substances from many sources, it is to be expected that its microbial flora will fluctuate both in types and numbers. Raw sewage may contain millions of bacteria per milliliter. Prominent among these are the coliforms, streptococci, anaerobic spore forming bacilli, the *Proteus* group, and other types which have their origin in the intestinal tract of man. Sewage is also a potential source of pathogenic intestinal organisms. The poliomyelitis virus has been demonstrated to occur in sewage; other viruses are readily isolated from the same source. Aside from the examination of sewage to demonstrate the presence of some specific microorganism for epidemiological purposes, bacteriological analysis provides little useful information because of the magnitude of variations known to occur with regard to both numbers and kinds.

20. According to the above paragraph,

    20.____

    A. all sewage contains pathogenic organisms
    B. bacteriological analysis of sewage is routinely performed in order to determine the presence of coliform organisms
    C. microorganisms found in sewage vary from time to time
    D. poliomyelitis epidemics are due to viruses found in sewage

21. The title which would be MOST suitable for the above paragraph is

    21.____

    A. DISPOSAL OF SEWAGE BY BACTERIA
    B. MICROBES AND SEWAGE TREATMENT
    C. MICROBIOLOGICAL CHARACTERISTICS OF SEWAGE
    D. SEWAGE REMOVAL PROCESSES

Questions 22-23.

DIRECTIONS: Questions 22 and 23 are to be answered SOLELY on the basis of the following paragraphs.

Most cities carrying on public health work exercise varying degrees of inspection and control over their milk supplies. In some cases, it consists only of ordinances, with little or no attempt at enforcement. In other cases, good control is obtained through wise ordinances and an efficient inspecting force and laboratory. While inspection alone can do much toward controlling the quality and production of milk, there must also be frequent laboratory tests of the milk.

The bacterial count of the milk indicates the condition of the dairy and the methods of milk handling. The counts, therefore, are a check on the reports of the sanitarian. High bacterial counts of milk from a dairy reported by a sanitarian to be *good* may indicate difficulty not suspected by the sanitarian such as infected udders, inefficient sterilization of utensils, or poor cooling.

22. According to the above passage, the MOST accurate of the following statements is: 22.___

    A. The bacterial count of milk will be low if milk-producing animals are free from disease.
    B. A high bacterial count of milk can be reduced by pasteurization.
    C. The bacterial count of milk can be controlled by the laboratory.
    D. The bacterial count of milk will be low if the conditions of milk production, processing, and handling are good.

23. The following conclusion may be drawn from the above passage: 23.___

    A. Large centers of urban population usually exercise complete control over their milk supplies
    B. Adequate legislation is an important adjunct of a milk supply control program
    C. Most cities should request the assistance of other cities prior to instituting a milk supply control program
    D. Wise laws establishing a milk supply control program obviate the need for the enforcement of such laws provided that good laboratory techniques are employed

Questions 24-25.

DIRECTIONS: Questions 24 and 25 are to be answered SOLELY on the basis of the following excerpt from the health code.

Article 101. Shellfish and Fish.

Section 101.03. Shippers of shellfish; registration.

    (a)   No shellfish shall be shipped into the city unless the shipper of such shellfish is registered with the department.
    (b)   Application for registration shall be made on a form furnished by the department.
    (c)   The following shippers shall be eligible to apply for registration:
        1.   A shipper of shellfish located in the state but outside the city who holds a shellfish shipper's permit issued by the state conservation department; or
        2.   A shipper of shellfish located outside the state, or located in Canada, who holds a shellfish certificate of approval or a permit issued by the state or provincial agency having control of the shellfish industry of his state or province, which certificate of approval or permit appears on the current list of interstate shellfish shipper permits published by the United States Public Health Service.
    (d)   The commissioner may refuse to accept the registration of any applicant whose past observance of the shellfish regulations is not satisfactory to the commissioner.
    (e)   No applicant shall ship shellfish into the city unless he has been notified in writing by the department that his application for registration has been approved.
    (f)   Every registration as a shipper of shellfish, unless sooner revoked, shall terminate on the expiration date of the registrant's state shellfish certificate or permit.

24. The above excerpt from the health code provides that

    A. permission to register may not be denied to a shellfish shipper meeting the standards of his own jurisdiction

    B. permission to register will not be denied unless the shipper's past observances of shellfish regulations has not been satisfactory to the U. S. Public Health Service

    C. the commissioner may suspend the regulations applicable to registration if requested to do so by the governmental agency having jurisdiction over the shellfish shipper

    D. an applicant for registration as a shellfish shipper may ship shellfish into the city when notified by the department in writing that his application has been approved

25. The above excerpt from the health code provides that

    A. applications for registration will not be granted to out-of-state shippers of shellfish who have already received permission to sell shellfish from another jurisdiction

    B. shippers of shellfish located outside of the city may not ship shellfish into the city unless the shellfish have passed inspection by the jurisdiction in which the shellfish shipper is located

    C. a shipper of shellfish located in Canada is eligible for registration provided that he holds a shellfish permit issued by the appropriate provincial agency and that such permit appears on the current list of shellfish shipper permits published by the U. S. Public Health Service

    D. a shipper of shellfish located in Canada whose shell-fish permit has been revoked by the provincial agency may ship shellfish into the city until such time as he is notified in writing by the department that his shellfish registration has been revoked

Questions 26-30.

DIRECTIONS: Questions 26 through 30 are to be answered SOLELY on the basis of the following passage.

Lots of exterminators still think accidents just happen – that they are due to bad luck. Nothing could be further from the truth. Evidence of this is in the drop in accidents among employees of the City of New York since the Safety Program started. The one-out-of-a-hundred accidents that cannot be prevented might be called *Acts of God.* They are things like lightning, earthquakes, tornadoes, and tidal waves that we are powerless to prevent – although we can take precautions against them which will cut down the accident rate. The other ninety-nine percent of the accidents clearly have a manmade cause. If you trace back far enough, you'll find that somewhere, somehow, someone could have done something to prevent these accidents. For just about every accident, there is some fellow who fouled up. He didn't protect himself, he didn't use the right equipment, he wasn't alert, he lost his temper, he didn't have his mind on his work, he was *kidding around,* or he took a shortcut because he was just too lazy. We must all work together to improve safety and prevent injury and death.

26. The one of the following titles which BEST describes the subject matter of the above passage is 26.____

    A. ACTS OF GOD
    B. THE IMPORTANCE OF SAFETY CONSCIOUSNESS
    C. SAFETY IN NEW YORK CITY
    D. WORKING TOGETHER

27. After New York City began to operate a safety program, it was found that 27.____

    A. the number of accidents were reduced
    B. production decreased
    C. accidents stayed the same but employees were more careful
    D. the element of bad luck did not change

28. One cause of accidents that is NOT mentioned in the above passage is 28.____

    A. failure to keep alert
    B. taking a shortcut
    C. using the wrong equipment
    D. working too fast

29. The number of accidents caused by such things as hurricanes can be 29.____

    A. changed only by an *Act of God*
    B. eliminated by strict adherence to safety rules
    C. increased by being too careful
    D. reduced by proper safety precautions

30. The percentage of accidents that occur as a result of things that CANNOT be prevented is approximately _____ percent. 30.____

    A. 1   B. 10   C. 50   D. 99

KEY (CORRECT ANSWERS)

1.	B		16.	D
2.	A		17.	D
3.	B		18.	D
4.	D		19.	D
5.	B		20.	C
6.	C		21.	C
7.	B		22.	D
8.	D		23.	B
9.	C		24.	D
10.	B		25.	C
11.	C		26.	B
12.	B		27.	A
13.	D		28.	D
14.	C		29.	D
15.	C		30.	A

SCIENCE READING COMPREHENSION
EXAMINATION SECTION
TEST 1

DIRECTIONS: Each question or incomplete statement is followed by several suggested answers or completions. Select the one that BEST answers the question or completes the statement. *PRINT THE LETTER OF THE CORRECT ANSWER IN THE SPACE AT THE RIGHT.*

PASSAGE

Photosynthesis is a complex process with many intermediate steps. Ideas differ greatly as to the details of these steps, but the general nature of the process and its outcome are well established. Water, usually from the soil, is conducted through the xylem of root, stem and leaf to the chlorophyl-containing cells of a leaf. In consequence of the abundance of water within the latter cells, their walls are saturated with water. Carbon dioxide, diffusing from the air through the stomata and into the intercellular spaces of the leaf, comes into contact with the water in the walls of the cells which adjoin the intercellular spaces. The carbon dioxide becomes dissolved in the water of these walls, and in solution diffuses through the walls and the plasma membranes into the cells. By the agency of chlorophyl in the chloroplasts of the cells, the energy of light is transformed into chemical energy. This chemical energy is used to decompose the carbon dioxide and water, and the products of their decomposition are recombined into a new compound. The compound first formed is successively built up into more and more complex substances until finally a sugar is produced.

Questions 1-8.

1. The union of carbon dioxide and water to form starch results in an excess of   1._____

    A. hydrogen   B. carbon   C. oxygen
    D. carbon monoxide   E. hydrogen peroxide

2. Synthesis of carbohydrates takes place   2._____

    A. in the stomata
    B. in the intercellular spaces of leaves
    C. in the walls of plant cells
    D. within the plasma membranes of plant cells
    E. within plant cells that contain chloroplasts

3. In the process of photosynthesis, chlorophyl acts as a   3._____

    A. carbohydrate   B. source of carbon dioxide
    C. catalyst   D. source of chemical energy
    E. plasma membrane

4. In which of the following places are there the GREATEST number of hours in which pho-   4._____
    tosynthesis can take place during the month of December?

    A. Buenos Aires, Argentina   B. Caracas, Venezuela
    C. Fairbanks, Alaska   D. Quito, Ecuador
    E. Calcutta, India

5. During photosynthesis, molecules of carbon dioxide enter the stomata of leaves because   5.___

    A. the molecules are already in motion
    B. they are forced through the stomata by the son's rays
    C. chlorophyl attracts them
    D. a chemical change takes place in the stomata
    E. oxygen passes out through the stomata

6. Besides food manufacture, another USEFUL result of photosynthesis is that it   6.___

    A. aids in removing poisonous gases from the air
    B. helps to maintain the existing proportion of gases in the air
    C. changes complex compounds into simpler compounds
    D. changes certain waste products into hydrocarbons
    E. changes chlorophyl into useful substances

7. A process that is almost the exact reverse of photosynthesis is the   7.___

    A. rusting of iron   B. burning of wood
    C. digestion of starch   D. ripening of fruit
    E. storage of food in seeds

8. The leaf of the tomato plant will be unable to carry on photosynthesis if the   8.___

    A. upper surface of the leaf is coated with vaseline
    B. upper surface of the leaf is coated with lampblack
    C. lower surface of the leaf is coated with lard
    D. leaf is placed in an atmosphere of pure carbon dioxide
    E. entire leaf is coated with lime

TEST 2

PASSAGE

The only carbohydrate which the human body can absorb and oxidize is the simple sugar glucose. Therefore, all carbohydrates which are consumed must be changed to glucose by the body before they can be used. There are specific enzymes in the mouth, the stomach, and the small intestine which break down complex carbohydrates. All the monosaccharides are changed to glucose by enzymes secreted by the intestinal glands, and the glucose is absorbed by the capillaries of the villi.

The following simple test is used to determine the presence of a reducing sugar. If Benedict's solution is added to a solution containing glucose or one of the other reducing sugars and the resulting mixture is heated, a brick-red precipitate will be formed. This test was carried out on several substances and the information in the following table was obtained. "P" indicates that the precipitate was formed and "N" indicates that no reaction was observed.

Material Tested	Observation
Crushed grapes in water	P
Cane sugar in water	N
Fructose	P
Molasses	N

Questions 1-2.

1. From the results of the test made upon crushed grapes in water, one may say that grapes contain

   A. glucose
   D. no sucrose
   B. sucrose
   E. no glucose
   C. a reducing sugar

1.____

2. Which one of the following foods probably undergoes the LEAST change during the process of carbohydrate digestion in the human body?

   A. Cane sugar
   D. Bread
   B. Fructose
   E. Potato
   C. Molasses

2.____

TEST 3

DIRECTIONS: Each question or incomplete statement is followed by several suggested answers or completions. Select the one that BEST answers the question or completes the statement. *PRINT THE LETTER OF THE CORRECT ANSWER IN THE SPACE AT THE RIGHT.*

PASSAGE

The British pressure suit was made in two pieces and joined around the middle in contrast to the other suits, which were one-piece suits with a removable helmet. Oxygen was supplied through a tube, and a container of soda lime absorbed carbon dioxide and water vapor. The pressure was adjusted to a maximum of 2 1/2 pounds per square inch (130 millimeters) higher than the surrounding air. Since pure oxygen was used, this produced a partial pressure of 130 millimeters, which is sufficient to sustain the flier at any altitude.

Using this pressure suit, the British established a world's altitude record of 49,944 feet in 1936 and succeeded in raising it to 53,937 feet the following year. The pressure suit is a compromise solution to the altitude problem. Full sea-level pressure can not be maintained, as the suit would be so rigid that the flier could not move arms or legs. Hence a pressure one third to one fifth that of sea level has been used. Because of these lower pressures, oxygen has been used to raise the partial pressure of alveolar oxygen to normal.

Questions 1-9.

1. The MAIN constituent of air not admitted to the pressure suit described was

    A. oxygen   B. nitrogen   C. water vapor
    D. carbon dioxide   E. hydrogen

1.____

2. The pressure within the suit exceeded that of the surrounding air by an amount equal to 130 millimeters of

    A. mercury   B. water   C. air
    D. oxygen   E. carbon dioxide

2.____

3. The normal atmospheric pressure at sea level is

    A. 130 mm   B. 250 mm   C. 760 mm
    D. 1000 mm   E. 1300 mm

3.____

4. The water vapor that was absorbed by the soda lime came from

    A. condensation
    B. the union of oxygen with carbon dioxide
    C. body metabolism
    D. the air within the pressure suit
    E. water particles in the upper air

4.____

5. The HIGHEST altitude that has been reached with the British pressure suit is about

    A. 130 miles   B. 2 1/2 miles   C. 6 miles
    D. 10 miles   E. 5 miles

5.____

6. If the pressure suit should develop a leak, the   6.____

    A. oxygen supply would be cut off
    B. suit would fill up with air instead of oxygen
    C. pressure within the suit would drop to zero
    D. pressure within the suit would drop to that of the surrounding air
    E. suit would become so rigid that the flier would be unable to move arms or legs

7. The reason why oxygen helmets are unsatisfactory for use in efforts to set higher altitude   7.____
records is that

    A. it is impossible to maintain a tight enough fit at the neck
    B. oxygen helmets are too heavy
    C. they do not conserve the heat of the body as pressure suits do
    D. if a parachute jump becomes necessary, it can not be made while such a helmet is being worn
    E. oxygen helmets are too rigid

8. The pressure suit is termed a compromise solution because   8.____

    A. it is not adequate for stratosphere flying
    B. aviators can not stand sea-level pressure at high altitudes
    C. some suits are made in two pieces, others in one
    D. other factors than maintenance of pressure have to be accommodated
    E. full atmospheric pressure can not be maintained at high altitudes

9. The passage implies that   9.____

    A. the air pressure at 49,944 feet is approximately the same as it is at 53,937 feet
    B. pressure cabin planes are not practical at extremely high altitudes
    C. a flier's oxygen requirement is approximately the same at high altitudes as it is at sea level
    D. one-piece pressure suits with removable helmets are unsafe
    E. a normal alveolar oxygen supply is maintained if the air pressure is between one third and one fifth that of sea level

———————

TEST 4

DIRECTIONS: Each question or incomplete statement is followed by several suggested answers or completions. Select the one that BEST answers the question or completes the statement. *PRINT THE LETTER OF THE CORRECT ANSWER IN THE SPACE AT THE RIGHT.*

PASSAGE

Chemical investigations show that during muscle contraction the store of organic phosphates in the muscle fibers is altered as energy is released. In doing so, the organic phosphates (chiefly adenoisine triphosphate and phospho-creatine) are transformed anaerobically to organic compounds plus phosphates. As soon as the organic phosphates begin to break down in muscle contraction, the glycogen in the muscle fibers also transforms into lactic acid plus free energy; this energy the muscle fiber uses to return the organic compounds plus phosphates into high-energy organic phosphates ready for another contraction. In the presence of oxygen, the lactic acid from the glycogen decomposition is changed also. About one-fifth of it is oxidized to form water and carbon dioxide and to yield another supply of energy. This time the energy is used to transform the remaining four-fifths of the lactic acid into glycogen again.

Questions 1-5.

1. The energy for muscle contraction comes directly from the   1.___

   A. breakdown of lactic acid into glycogen
   B. resynthesis of adenosine triphosphate
   C. breakdown of glycogen into lactic acid
   D. oxidation of lactic acid
   E. breakdown of the organic phosphates

2. Lactic acid does NOT accumulate in a muscle that   2.___

   A. is in a state of lacking oxygen
   B. has an ample supply of oxygen
   C. is in a state of fatigue
   D. is repeatedly being stimulated
   E. has an ample supply of glycogen

3. The energy for the resynthesis of adenosine triphosphate and phospho-creatine comes   3.___
   from the

   A. oxidation of lactic acid
   B. synthesis of organic phosphates
   C. change from glycogen to lactic acid
   D. resynthesis of glycogen
   E. change from lactic acid to glycogen

4. The energy for the resynthesis of glycogen comes from the   4.___

   A. breakdown of organic phosphates
   B. resynthesis of organic phosphates
   C. change occurring in one-fifth of the lactic acid

D. change occurring in four-fifths of the lactic acid
E. change occurring in four-fifths of glycogen

5. The breakdown of the organic phosphates into organic compounds plus phosphates is an   5._____

A. anobolic reaction   B. aerobic reaction
C. endothermic reaction   D. exothermic reaction
E. anaerobic reaction

———————

TEST 5

PASSAGE

And with respect to that theory of the origin of the forms of life peopling our globe, with which Darwin's name is bound up as closely as that of Newton with the theory of gravitation, nothing seems to be further from the mind of the present generation than any attempt to smother it with ridicule or to crush it by vehemence of denunciation. "The struggle for existence," and "natural selection," have become household words and everyday conceptions. The reality and the importance of the natural processes on which Darwin founds his deductions are no more doubted than those of growth and multiplication; and, whether the full potency attributed to them is admitted or not, no one is unmindful of or at all doubts their vast and far-reaching significance. Wherever the biological sciences are studied, the "Origin of Species" lights the path of the investigator; wherever they are taught it permeates the course of instruction. Nor has the influence of Darwinian ideas been less profound beyond the realms of biology. The oldest of all philosophies, that of evolution, was bound hand and foot and cast into utter darkness during the millennium of theological scholasticism. But Darwin poured new life-blood into the ancient frame; the bonds burst, and the revivified thought of ancient Greece has proved itself to be a more adequate expression of the universal order of things than any of the schemes which have been accepted by the credulity and welcomed by the superstition of seventy later generations of men.

Questions 1-7.

1. Darwin's theory of the origin of the species is based on   1._____

    A. theological deductions
    B. the theory of gravitation
    C. Greek mythology
    D. natural processes evident in the universe
    E. extensive reading in the biological sciences

2. The passage implies that   2._____

    A. thought in ancient Greece was dead
    B. the theory of evolution is now universally accepted
    C. the "Origin of Species" was seized by the Church
    D. Darwin was influenced by Newton
    E. the theories of "the struggle for existence" and "natural selection" are too evident to be scientific

3. The idea of evolution   3._____

    A. was suppressed for 1,000 years
    B. is falsely claimed by Darwin
    C. has swept aside all superstition
    D. was outworn even in ancient Greece
    E. has revolutionized the universe

4. The processes of growth and multiplication   4._____

    A. have been replaced by others discovered by Darwin
    B. were the basis for the theory of gravitation
    C. are "the struggle for existence" and "natural selection"
    D. are scientific theories not yet proved
    E. are accepted as fundamental processes of nature

5. Darwin's treatise on evolution   5._____

    A. traces life on the planets from the beginning of time to the present day
    B. was translated from the Greek
    C. contains an ancient philosophy in modern, scientific guise
    D. has had a profound effect on evolution
    E. has had little notice outside scientific circles

6. The theory of evolution   6._____

    A. was first advanced in the "Origin of Species"
    B. was suppressed by the ancient Greeks
    C. did not get beyond the monasteries during the millennium
    D. is philosophical, not scientific
    E. was elaborated and revived by Darwin

7. Darwin has contributed GREATLY toward   7._____

    A. a universal acceptance of the processes of nature
    B. reviving the Greek intellect
    C. ending the millennium of theological scholasticism
    D. a satisfactory explanation of scientific theory
    E. easing the struggle for existence

TEST 6

DIRECTIONS: Each question or incomplete statement is followed by several suggested answers or completions. Select the one that BEST answers the question or completes the statement. *PRINT THE LETTER OF THE CORRECT ANSWER IN THE SPACE AT THE RIGHT.*

PASSAGE

The higher forms of plants and animals, such as seed plants and vertebrates, are similar or alike in many respects but decidedly different in others. For example, both of these groups of organisms carry on digestion, respiration, reproduction, conduction, growth, and exhibit sensitivity to various stimuli. On the other hand, a number of basic differences are evident. Plants have no excretory systems comparable to those of animals. Plants have no heart or similar pumping organ. Plants are very limited in their movements. Plants have nothing similar to the animal nervous system. In addition, animals can not synthesize carbohydrates from inorganic substances. Animals do not have special regions of growth, comparable to terminal and lateral meristems in plants, which persist through-out the life span of the organism. And, finally, the animal cell "wall" is only a membrane, while plant cell walls are more rigid, usually thicker, and may be composed of such substances as cellulose, lignin, pectin, cutin, and suberin. These characteristics are important to an understanding of living organisms and their functions and should, consequently, be carefully considered in plant and animal studies

Questions 1-7.

1. Which of the following do animals lack?   1._____

    A. Ability to react to stimuli
    B. Ability to conduct substances from one place to another
    C. Reproduction by gametes
    D. A cell membrane
    E. A terminal growth region

2. Which of the following statements is false?   2._____

    A. Animal cell "walls" are composed of cellulose.
    B. Plants grow as long as they live.
    C. Plants produce sperms and eggs.
    D. All vertebrates have hearts.
    E. Wood is dead at maturity.

3. Respiration in plants takes place   3._____

    A. only during the day
    B. only in the presence of carbon dioxide
    C. both day and night
    D. only at night
    E. only in the presence of certain stimuli

4. An example of a vertebrate is the   4._____

    A. earthworm   B. starfish   C. amoeba
    D. cow   E. insect

5. Which of the following statements is true?   5.____

    A.   All animals eat plants as a source of food.
    B.   Respiration, in many ways, is the reverse of photo-synthesis.
    C.   Man is an invertebrate animal.
    D.   Since plants have no hearts, they can not develop high pressures in their cells.
    E.   Plants can not move.

6. Which of the following do plants lack?   6.____

    A.   A means of movement
    B.   Pumping structures
    C.   Special regions of growth
    D.   Reproduction by gametes
    E.   A digestive process

7. A substance that can be synthesized by green plants but NOT by animals is   7.____

    A.   protein   B.   cellulose   C.   carbon dioxide
    D.   uric acid   E.   water

————

TEST 7

PASSAGE

Sodium chloride, being by far the largest constituent of the mineral matter of the blood, assumes special significance in the regulation of water exchanges in the organism. And, as Cannon has emphasized repeatedly, these latter are more extensive and more important than may at first thought appear. He points out "there are a number of circulations of the fluid out of the body and back again, without loss." Thus, by example, it is estimated that from a quart and one-half of water daily "leaves the body" when it enters the mouth as saliva; another one or two quarts are passed out as gastric juice; and perhaps the same amount is contained in the bile and the secretions of the pancreas and the intestinal wall. This large volume of water enters the digestive processes; and practically all of it is reabsorbed through the intestinal wall, where it performs the equally important function of carrying in the digested foodstuffs. These and other instances of what Cannon calls "the conservative use of water in our bodies" involve essentially osmotic pressure relationships in which the concentration of sodium chloride plays an important part.

Questions 1-11.

1. This passage implies that

    A. the contents of the alimentary canal are not to be considered within the body
    B. sodium chloride does not actually enter the body
    C. every particle of water ingested is used over and over again
    D. water can not be absorbed by the body unless it contains sodium chloride
    E. substances can pass through the intestinal wall in only one direction

2. According to this passage, which of the following processes requires MOST water? The

    A. absorption of digested foods
    B. secretion of gastric juice
    C. secretion of saliva
    D. production of bile
    E. concentration of sodium chloride solution

3. A body fluid that is NOT saline is

    A. blood   B. urine   C. bile
    D. gastric juice   E. saliva

4. An organ that functions as a storage reservoir from which large quantities of water are reabsorbed into the body is the

    A. kidney   B. liver   C. large intestine
    D. mouth   E. pancreas

5. Water is reabsorbed into the body by the process of   5.____

   A. secretion   B. excretion   C. digestion
   D. osmosis   E. oxidation

6. Digested food enters the body PRINCIPALLY through the   6.____

   A. mouth   B. liver   C. villi
   D. pancreas   E. stomach

7. The metallic element found in the blood in compound form and present there in larger   7.____
   quantities than any other metallic element is

   A. iron   B. calcium   C. magnesium
   D. chlorine   E. sodium

8. An organ that removes water from the body and prevents its reabsorption for use in the 8.____
   body processes is the

   A. pancreas   B. liver   C. small intestine
   D. lungs   E. large intestine

9. In which of the following processes is sodium chloride removed MOST rapidly from the   9.____
   body?

   A. Digestion   B. Breathing   C. Oxidation
   D. Respiration   E. Perspiration

10. Which of the following liquids would pass from the alimentary canal into the blood MOST 10.____
    rapidly?

    A. A dilute solution of sodium chloride in water
    B. Gastric juice
    C. A concentrated solution of sodium chloride in water
    D. Digested food
    E. Distilled water

11. The reason why it is unsafe to drink ocean water even under conditions of extreme thirst 11.____
    is that it

    A. would reduce the salinity of the blood to a dangerous level
    B. contains dangerous disease germs
    C. contains poisonous salts
    D. would greatly increase the salinity of the blood
    E. would cause salt crystals to form in the blood stream

_____

TEST 8

DIRECTIONS: Each question or incomplete statement is followed by several suggested answers or completions. Select the one that BEST answers the question or completes the statement. *PRINT THE LETTER OF THE CORRECT ANSWER IN THE SPACE AT THE RIGHT.*

PASSAGE

The discovery of antitoxin and its specific antagonistic effect upon toxin furnished an opportunity for the accurate investigation of the relationship of a bacterial antigen and its antibody. Toxin-antitoxin reactions were the first immunological processes to which experimental precision could be applied, and the discovery of principles of great importance resulted from such studies. A great deal of the work was done with diphtheria toxin and antitoxin and the facts elucidated with these materials are in principle applicable to similar substances.

The simplest assumption to account for the manner in which an antitoxin renders a toxin innocuous would be that the antitoxin destroys the toxin. Roux and Buchner, however, advanced the opinion that the antitoxin did not act directly upon the toxin, but affected it indirectly through the mediation of tissue cells. Ehrlich, on the other hand, conceived the reaction of toxin and antitoxin as a direct union, analogous to the chemical neutralization of an acid by a base.

The conception of toxin destruction was conclusively refuted by the experiments of Calmette. This observer, working with snake poison, found that the poison itself (unlike most other toxins) possessed the property of resisting heat to 100 degrees C, while its specific antitoxin, like other antitoxins, was destroyed at or about 70 degrees C. Nontoxic mixtures of the two substanues, when subjected to heat, regained their toxic properties. The natural inference from these observations was that the toxin in the original mixture had not been destroyed, but had been merely inactiviated by the presence of the antitoxin and again set free after destruction of the antitoxin by heat.

Questions 1-10.

1. Both toxins and antitoxins ORDINARILY   1.____

    A.   are completely destroyed at body temperatures
    B.   are extremely resistant to heat
    C.   can exist only in combination
    D.   are destroyed at 180° F
    E.   are products of nonliving processes

2. MOST toxins can be destroyed by   2.____

    A.   bacterial action   B.   salt solutions
    C.   boiling   D.   diphtheria antitoxin
    E.   other toxins

3. Very few disease organisms release a true toxin into the blood stream. It would follow,   3.____
then, that

    A.   studies of snake venom reactions have no value
    B.   studies of toxin-antitoxin reactions are of little importance

    C.   the treatment of most diseases must depend upon information obtained from study of a few

    D.   antitoxin plays an important part in the body defense against the great majority of germs

    E.   only toxin producers are dangerous

4.   A person becomes susceptible to infection again immediately after recovering from   4._____

    A. mumps   B. tetanus   C. diphtheria
    D. smallpox   E. tuberculosis

5.   City people are more frequently immune to communicable diseases than country people are because   5._____

    A.   country people eat better food
    B.   city doctors are better than country doctors
    C.   the air is more healthful in the country
    D.   country people have fewer contacts with disease carriers
    E.   there are more doctors in the city than in the country

6.   The substances that provide us with immunity to disease are found in the body in the   6._____

    A. blood serum   B. gastric juice   C. urine
    D. white blood cells   E. red blood cells

7.   A person ill with diphtheria would MOST likely be treated with   7._____

    A. diphtheria toxin   B. diphtheria toxoid
    C. dead diphtheria germs   D. diphtheria antitoxin
    E. live diphtheria germs

8.   To determine susceptibility to diphtheria, an individual may be given the   8._____

    A. Wassermann test   B. Schick test
    C. Widal test   D. Dick test
    E. Kahn test

9.   Since few babies under six months of age contract diphtheria, young babies PROBABLY   9._____

    A.   are never exposed to diphtheria germs
    B.   have high body temperatures that destroy the toxin if acquired
    C.   acquire immunity from their mothers
    D.   acquire immunity from their fathers
    E.   are too young to become infected

10.   Calmette's findings   10._____

    A.   contradicted both Roux and Buchner's opinion and Ehrlich's conception
    B.   contradicted Roux and Buchner, but supported Ehrlich
    C.   contradicted Ehrlich, but supported Roux and Buchner
    D.   were consistent with both theories
    E.   had no bearing on the point at issue

TEST 9

DIRECTIONS: Each question or incomplete statement is followed by several suggested answers or completions. Select the one that BEST answers the question or completes the statement. *PRINT THE LETTER OF THE CORRECT ANSWER IN THE SPACE AT THE RIGHT.*

PASSAGE

In the days of sailing ships, when voyages were long and uncertain, provisions for many months were stored without refrigeration in the holds of the ships. Naturally no fresh or perishable foods could be included. Toward the end of particularly long voyages the crews of such ships became ill and often many died from scurvy. Many men, both scientific and otherwise, tried to devise a cure for scurvy. Among the latter was John Hall, a son-in-law of William Shakespeare, who cured some cases of scurvy by administering a sour brew made from scurvy grass and water cress.

The next step was the suggestion of William Harvey that scurvy could be prevented by giving the men lemon juice. He thought that the beneficial substance was the acid contained in the fruit.

The third step was taken by Dr. James Lind, an English naval surgeon, who performed the following experiment with 12 sailors, all of whom were sick with scurvy: Each was given the same diet, except that four of the men received small amounts of dilute sulfuric acid, four others were given vinegar and the remaining four were given lemons. Only those who received the fruit recovered.

Questions 1-7.

1. Credit for solving the problem described above belongs to   1._____

    A. Hall, because he first devised a cure for scurvy
    B. Harvey, because he first proposed a solution of the problem
    C. Lind, because he proved the solution by means of an experiment
    D. both Harvey and Lind, because they found that lemons are more effective than scurvy grass or water cress
    E. all three men, because each made some contribution

2. A good substitute for lemons in the treatment of scurvy is   2._____

    A. fresh eggs   B. tomato juice   C. cod-liver oil
    D. liver   E. whole-wheat bread

3. The number of control groups that Dr. Lind used in his experiment was   3._____

    A. one   B. two   C. three   D. four   E. none

4. A substance that will turn blue litmus red is   4._____

    A. aniline   B. lye   C. ice
    D. vinegar   E. table salt

5. The hypothesis tested by Lind was: 5._____

    A. Lemons contain some substance not present in vinegar.
    B. Citric acid is the most effective treatment for scurvy.

C. Lemons contain some unknown acid that will cure scurvy.
D. Some specific substance, rather than acids in general, is needed to cure scurvy.
E. The substance needed to cure scurvy is found only in lemons.

6. A problem that Lind's experiment did NOT solve was: 6.____

A. Will citric acid alone cure scurvy?
B. Will lemons cure scurvy?
C. Will either sulfuric acid or vinegar cure scurvy?
D. Are all substances that contain acids equally effective as a treatment for scurvy?
E. Are lemons more effective than either vinegar or sulfuric acid in the treatment of scurvy?

7. The PRIMARY purpose of a controlled scientific experiment is to   7.____

A. get rid of superstitions
B. prove a hypothesis is correct
C. disprove a theory that is false
D. determine whether a hypothesis is true or false
E. discover new facts

TEST 10

PASSAGE

The formed elements of the blood are the red corpuscles or erythrocytes, the white corpuscles or leucocytes, the blood platelets, and the so-called blood dust or hemoconiae. Together, these constitute 30-40 per cent by volume of the whole blood, the remainder being taken up by the plasma. In man, there are normally 5,000,000 red cells per cubic millimeter of blood; the count is somewhat lower in women. Variations occur frequently, especially after exercise or a heavy meal, or at high altitudes. Except in camels, which have elliptical corpuscles, the shape of the mammalian corpuscle is that of a circular, nonnucleated, bi-concave disk. The average diameter usually given is 7.7 microns, a value obtained by examining dried preparations of blood and considered by Ponder to be too low. Ponder's own observations, made on red cells in the fresh state, show the human corpuscle to have an average diameter of 8.8 microns. When circulating in the blood vessels, the red cell does not maintain a fixed shape but changes its form constantly, especially in the small capillaries. The red blood corpuscles are continually undergoing destruction, new corpuscles being formed to replace them. The average life of red corpuscles has been estimated by various investigators to be between three and six weeks. Preceding destruction, changes in the composition of the cells are believed to occur which render them less resistant. In the process of destruction, the lipids of the membrane are dissolved and the hemoglobin which is liberated is the most important, though probably not the only, source of bilirubin. The belief that the liver is the only site of red cell destruction is no longer generally held. The leucocytes, of which there are several forms, usually number between 7000 and 9000 per cubic millimeter of blood. These increase in number in disease, particularly when there is bacterial infection.

Questions 1-10.

1. Leukemia is a disease involving the   1.___

    A. red cells   B. white cells   C. plasma
    D. blood platelets   E. blood dust

2. Are the erythrocytes in the blood increased in number after a heavy meal? The paragraph implies that this   2.___

    A. is true   B. holds only for camels
    C. is not true   D. may be true
    E. depends on the number of white cells

3. When blood is dried, the red cells   3.___

    A. contract   B. remain the same size   C. disintegrate
    D. expand   E. become elliptical

4. Ponder is probably classified as a professional   4.___

    A. pharmacist   B. physicist   C. psychologist
    D. physiologist   E. psychiatrist

5. The term "erythema" when applied to skin conditions signifies   5.____

    A. redness   B. swelling   C. irritation
    D. pain   E. roughness

6. Lipids are insoluble in water and soluble in such solvents as ether, chloroform and ben-   6.____
zene. It may be inferred that the membranes of red cells MOST closely resemble

    A. egg white   B. sugar   C. bone
    D. butter   E. cotton fiber

7. Analysis of a sample of blood yields cell counts of 4,800,000 erythrocytes and 16,000   7.____
leucocytes per cubic millimeter. These data suggest that the patient from whom the
blood was taken

    A. is anemic
    B. has been injuriously invaded by germs
    C. has been exposed to high-pressure air
    D. has a normal cell count
    E. has lost a great deal of blood

8. Bilirubin, a bile pigment, is   8.____

    A. an end product of several different reactions
    B. formed only in the liver
    C. formed from the remnants of the cell membranes of erythrocytes
    D. derived from hemoglobin exclusively
    E. a precursor of hemoglobin

9. Bancroft found that the blood count of the natives in the Peruvian Andes differed from   9.____
that usually accepted as normal. The blood PROBABLY differed in respect to

    A. leucocytes   B. blood platelets   C. cell shapes
    D. erythrocytes   E. hemoconiae

10. Hemoglobin is probably NEVER found   10.____

    A. free in the blood stream
    B. in the red cells
    C. in women's blood
    D. in the blood after exercise
    E. in the leucocytes

TEST 11

Questions 1-7.

DIRECTIONS: Each question or incomplete statement is followed by several suggested answers or completions. Select the one that BEST answers the question or completes the statement. *PRINT THE LETTER OF THE CORRECT ANSWER IN THE SPACE AT THE RIGHT.*

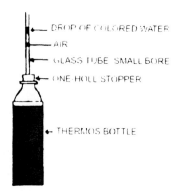

DROP OF COLORED WATER
AIR
GLASS TUBE SMALL BORE
ONE-HOLL STOPPER

THERMOS BOTTLE

1. The device shown in the diagram above indicates changes that are measured more accurately by a(n)

    A. thermometer   B. hygrometer   C. anemometer
    D. hydrometer   E. barometer

1._____

2. If the device is placed in a cold refrigerator for 72 hours, which of the following is MOST likely to happen?

    A. The stopper will be forced out of the bottle.
    B. The drop of water will evaporate.
    C. The drop will move downward.
    D. The drop will move upward.
    E. No change will take place.

2._____

3. When the device was carried in an elevator from the first floor to the sixth floor of a build-ing, the drop of colored water moved about 1/4 inch in the tube. Which of the following is MOST probably true? The drop moved

    A. *downward* because there was a decrease in the air pressure
    B. *upward* because there was a decrease in the air pressure
    C. *downward* because there was an increase in the air temperature
    D. *upward* because there was an increase in the air temperature
    E. *downward* because there was an increase in the temperature and a decrease in the pressure

3._____

4. The part of a thermos bottle into which liquids are poured consists of

    A. a single-walled, metal flask coated with silver
    B. two flasks, one of glass and one of silvered metal
    C. two silvered-glass flasks separated by a vacuum
    D. two silver flasks separated by a vacuum
    E. a single-walled, glass flask with a silver-colored coating

4._____

5. The thermos bottle is MOST similar in principle to   5.____

    A.   the freezing unit in an electric refrigerator
    B.   radiant heaters
    C.   solar heating systems
    D.   storm windows
    E.   a thermostatically controlled heating system

6. In a plane flying at an altitude where the air pressure is only half the normal pressure at   6.____
sea level, the plane's altimeter should read, *approximately,*

    A.   3000 feet   B.   9000 feet   C.   18000 feet
    D.   27000 feet   E.   60000 feet

7. Which of the following is the POOREST conductor of heat?   7.____

    A.   Air under a pressure of 1.5 pounds per square inch
    B.   Air under a pressure of 15 pounds per square inch
    C.   Unsilvered glass
    D.   Silvered glass
    E.   Silver

TEST 12

DIRECTIONS: Each question or incomplete statement is followed by several suggested
answers or completions. Select the one that BEST answers the question or
completes the statement. *PRINT THE LETTER OF THE CORRECT ANSWER
IN THE SPACE AT THE RIGHT.*

PASSAGE

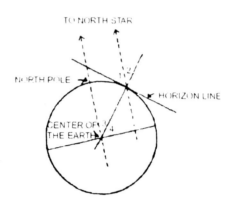

The latitude of any point on the earth's surface is the angle between a plumb line
dropped to the center of the earth from that point and the plane of the earth's equator. Since
it is impossible to go to the center of the earth to measure latitude, the latitude of any point
may be determined indirectly as shown in the accompanying dia gram.

It will be recalled that the axis of the earth, if extended out-ward, passes very near the
North Star. Since the North Star is, for all practical purposes, infinitely distant, the line of sight
to the North Star of an observer on the surface of the earth is virtually parallel with the earth's
axis. Angle 1, then, in the diagram represents the angular distance of the North Star above
the horizon. Angle 2 is equal to angle 3, because when two parallel lines are intersected by a
straight line, the corresponding angles are equal. Angle 1 plus angle 2 is a right angle and so
is angle 3 plus angle 4. Therefore, angle 1 equals angle 4 because when equals are sub-
tracted from equals the results are equal.

Questions 1-10.

1.   If an observer finds that the angular distance of the North Star above the horizon is 30,   1.____
his latitude is

   A.   15° N   B.   30° N   C.   60° N   D.   90° N   E.   120° N

2.   To an observer on the equator, the North Star would be   2.____

   A.   30° above the horizon   B.   60° above the horizon
   C.   90° above the horizon   D.   on the horizon
   E.   below the horizon

3. To an observer on the Arctic Circle, the North Star would be   3.____

    A. directly overhead
    B. 23 1/2° above the horizon
    C. 66 1/2° above the horizon
    D. on the horizon
    E. below the horizon

4. The distance around the earth along a certain parallel of latitude is 3600 miles. At that latitude, how many miles are there in one degree of longitude?   4.____

    A. 1 mile   B. 10 miles   C. 30 miles
    D. 69 miles   E. 100 miles

5. At which of the following latitudes would the sun be DIRECTLY overhead at noon on June 21?   5.____

    A. 0°   B. 23 1/2°S   C. 23 1/2°N
    D. 66 1/2°N   E. 66 1/2°S

6. On March 21 the number of hours of daylight at places on the Arctic Circle is   6.____

    A. none   B. 8   C. 12   D. 16   E. 24

7. The distance from the equator to the 45th parallel, measured along a meridian, is, *approximately,*   7.____

    A. 450 miles   B. 900 miles   C. 1250 miles
    D. 3125 miles   E. 6250 miles

8. The difference in time between the meridians that pass through longitude 45°E and longitude 105°W   8.____

    A. 6 hours   B. 2 hours   C. 8 hours
    D. 4 hours   E. 10 hours

9. Which of the following is NOT a great circle or part of a great circle?   9.____

    A. Arctic Circle
    B. 100th meridian
    C. Equator
    D. Shortest distance between New York and London
    E. Greenwich meridian

10. At which of the following places does the sun set EARLIEST on June 21?   10.____

    A. Montreal, Canada   B. Santiago, Chile
    C. Mexico City, Mexico   D. Lima, Peru
    E. Manila, P.I.

KEY (CORRECT ANSWERS)

TEST 1

1. C   5. A
2. E   6. B
3. C   7. B
4. A   8. C

TEST 2

1. C
2. B

TEST 3

1. B   6. D
2. A   7. D
3. C   8. E
4. C   9. C
5. D

TEST 4

1. A
2. B
3. C
4. C
5. D

TEST 5

1. D   5. D
2. B   6. E
3. A   7. A
4. E

TEST 6

1. E   5. B
2. A   6. B
3. C   7. B
4. D

TEST 7

1. A   6. C
2. A   7. E
3. D   8. D
4. C   9. E
5. D   10. E

TEST 8

1. D   6. A
2. C   7. D
3. C   8. B
4. E   9. C
5. D   10. D

TEST 9

1. E   5. D
2. B   6. A
3. B   7. D
4. D

TEST 10

1. B   6. D
2. D   7. B
3. A   8. A
4. D   9. D
5. A   10. E

TEST 11

1. A   5. D
2. C   6. C
3. B   7. A
4. C

TEST 12

1. B   6. C
2. D   7. D
3. C   8. E
4. B   9. A
5. C   10. B

EVALUATING INFORMATION AND EVIDENCE
EXAMINATION SECTION
TEST 1

DIRECTIONS: Each question or incomplete statement is followed by several suggested answers or completions. Select the one that BEST answers the question or completes the statement. *PRINT THE LETTER OF THE CORRECT ANSWER IN THE SPACE AT THE RIGHT.*

Questions 1 -9

Questions 1 through 9 measure your ability to (1) determine whether statements from witnesses say essentially the same thing and (2) determine the evidence needed to make it reasonably certain that a particular conclusion is true.

1. Which of the following pairs of statements say essentially the same thing in two different ways?　　　　1.＿＿＿

    I.   If you get your feet wet, you will catch a cold.
         If you catch a cold, you must have gotten your feet wet.
    II. If I am nominated, I will run for office.
         I will run for office only if I am nominated.

    A.   I only
    B.   I and II
    C.   II only
    D.   Neither I nor II

2. Which of the following pairs of statements say essentially the same thing in two different ways?　　　　2.＿＿＿

    I.   The enzyme Rhopsin cannot be present if the bacterium Trilox is absent.
         Rhopsin and Trilox always appear together.
    II. A member of PENSA has an IQ of at least 175.
         A person with an IQ of less than 175 is not a member of PENSA.

    A.   I only
    B.   I and II
    C.   II only
    D.   Neither I nor II

3. Which of the following pairs of statements say essentially the same thing in two different ways?　　　　3.＿＿＿

    I.   None of Finer High School's sophomores will be going to the prom.
         No student at Finer High School who is going to the prom is a sophomore.
    II. If you have 20/20 vision, you may carry a firearm.
         You may not carry a firearm unless you have 20/20 vision.

    A.   I only
    B.   I and II
    C.   II only
    D.   Neither I nor II

4. Which of the following pairs of statements say essentially the same thing in two different ways?

      I. If the family doesn't pay the ransom, they will never see their son again.
        It is necessary for the family to pay the ransom in order for them to see their son again.
      II. If it is raining, I am carrying an umbrella.
        If I am carrying an umbrella, it is raining.

    A. I only
    B. I and II
    C. II only
    D. Neither I nor II

4.____

5. <u>Summary of Evidence Collected to Date:</u>
In the county's maternity wards, over the past year, only one baby was born who did not share a birthday with any other baby.
<u>Prematurely Drawn Conclusion:</u> At least one baby was born on the same day as another baby in the county's maternity wards.
Which of the following pieces of evidence, if any, would make it *reasonably certain* that the conclusion drawn is true?

    A. More than 365 babies were born in the county's maternity wards over the past year
    B. No pairs of twins were born over the past year in the county's maternity wards
    C. More than one baby was born in the county's maternity wards over the past year
    D. None of these

5.____

6. <u>Summary of Evidence Collected to Date:</u>
Every claims adjustor for MetroLife drives only a Ford sedan when on the job.
<u>Prematurely Drawn Conclusion:</u> A person who works for MetroLife and drives a Ford sedan is a claims adjustor.
Which of the following pieces of evidence, if any, would make it *reasonably certain* that the conclusion drawn is true?

    A. Most people who work for MetroLife are claims adjustors
    B. Some people who work for MetroLife are not claims adjustors
    C. Most people who work for MetroLife drive Ford sedans
    D. None of these

6.____

7. <u>Summary of Evidence Collected to Date:</u>
Mason will speak to Zisk if Zisk will speak to Ronaldson.
<u>Prematurely Drawn Conclusion:</u> Jones will not speak to Zisk if Zisk will speak to Ronaldson
Which of the following pieces of evidence, if any, would make it *reasonably certain* that the conclusion drawn is true?

    A. If Zisk will speak to Mason, then Ronaldson will not speak to Jones
    B. If Mason will speak to Zisk, then Jones will not speak to Zisk
    C. If Ronaldson will speak to Jones, then Jones will speak to Ronaldson
    D. None of these

7.____

8. Summary of Evidence Collected to Date:
    No blue lights on the machine are indicators for the belt drive status.
    Prematurely Drawn Conclusion: Some of the lights on the lower panel are not indicators for the belt drive status.
    Which of the following pieces of evidence, if any, would make it *reasonably certain* that the conclusion drawn is true?

    A. No lights on the machine's lower panel are blue
    B. An indicator light for the machine's belt drive status is either green or red
    C. Some lights on the machine's lower panel are blue
    D. None of these

8.____

9. Summary of Evidence Collected to Date:
    Of the four Sweeney sisters, two are married, three have brown eyes, and three are doctors.
    Prematurely Drawn Conclusion: Two of the Sweeney sisters are brown-eyed, married doctors.
    Which of the following pieces of evidence, if any, would make it *reasonably certain* that the conclusion drawn is true?

    A. The sister who does not have brown eyes is married
    B. The sister who does not have brown eyes is not a doctor, and one who is not married is not a doctor
    C. Every Sweeney sister with brown eyes is a doctor
    D. None of these

9.____

Questions 10-14

Questions 10 through 14 refer to Map #5 and measure your ability to orient yourself within a given section of town, neighborhood or particular area. Each of the questions describes a starting point and a destination. Assume that you are driving a car in the area shown on the map accompanying the questions. Use the map as a basis for the shortest way to get from one point to another without breaking the law.

On the map, a street marked by arrows, or by arrows and the words "One Way," indicates one-way travel, and should be assumed to be one-way for the entire length, even when there are breaks or jogs in the street. EXCEPTION: A street that does not have the same name over the full length.

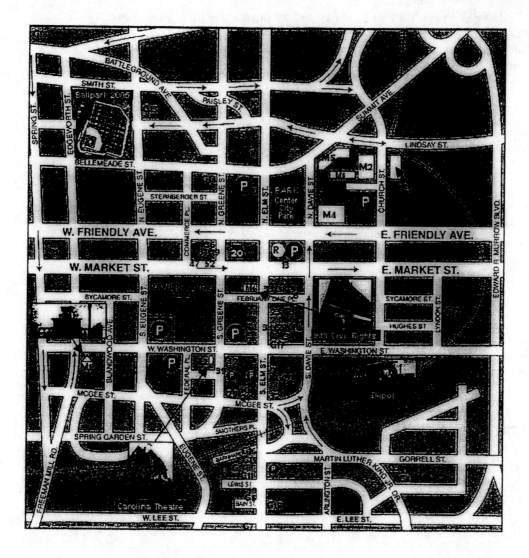

Map #5

10. The shortest legal way from the depot to Center City Park is

    A. north on Church, west on Market, north on Elm
    B. east on Washington, north on Edward R. Murrow Blvd., west on Friendly Ave.
    C. west on Washington, north on Greene, east on Market, north on Davie
    D. north on Church, west on Friendly Ave.

10.____

11. The shortest legal way from the Governmental Plaza to the ballpark is

    A. west on Market, north on Edgeworth
    B. west on Market, north on Eugene
    C. north on Greene, west on Lindsay
    D. north on Commerce Place, west on Bellemeade

11.____

12. The shortest legal way from the International Civil Rights Building to the building marked "M3" on the map is   12.____

    A. east on February One Place, north on Davie, east on Friendly Ave., north on Church
    B. south on Elm, west on Washington, north on Greene, east on Market, north on Church
    C. north on Elm, east on Market, north on Church
    D. north on Elm, east on Lindsay, south on Church

13. The shortest legal way from the ballpark to the Carolina Theatre is   13.____

    A. east on Lindsay, south on Greene
    B. south on Edgeworth, east on Friendly Ave., south on Greene
    C. east on Bellemeade, south on Elm, west on Washington
    D. south on Eugene, east on Washington

14. A car traveling north or south on Church Street may NOT go   14.____

    A. west onto Friendly Ave.
    B. west onto Lindsay
    C. east onto Market
    D. west onto Smith

Questions 15-19

Questions 15 through 19 refer to Figure #5, on the following page, and measure your ability to understand written descriptions of events. Each question presents a description of an accident or event and asks you which of the five drawings in Figure #5 BEST represents it.

In the drawings, the following symbols are used:

Moving vehicle: ⬆   Non-moving vehicle: ⬆

Pedestrian or bicyclist: ●

The path and direction of travel of a vehicle or pedestrian is indicated by a solid line.

The path and direction of travel of each vehicle or pedestrian directly involved in a collision from the point of impact is indicated by a dotted line.

In the space at the right, print the letter of the drawing that best fits the descriptions written below:

15. A driver heading south on Ohio runs a red light and strikes the front of a car headed west on Grand. He glances off and leaves the roadway at the southwest corner of Grand and Ohio.   15.____

16. A driver heading east on Grand drifts into the oncoming lane as it travels through the intersection of Grand and Ohio, and strikes an oncoming car head-on.   16.____

17. A driver heading east on Grand veers into the oncoming lane, sideswipes a westbound car and overcorrects as he swerves back into his lane. He leaves the roadway near the southeast corner of Grand and Ohio.

17.____

18. A driver heading east on Grand strikes the front of a car that is traveling north on Ohio and has run a red light. After striking the front of the northbound car, the driver veers left and leaves the roadway at the northeast corner of Grand and Ohio.

18.____

19. A driver heading east on Grand is traveling above the speed limit and clips the rear end of another eastbound car. The driver then veers to the left and leaves the roadway at the northeast corner of Grand and Ohio.

19.____

FIGURE #5

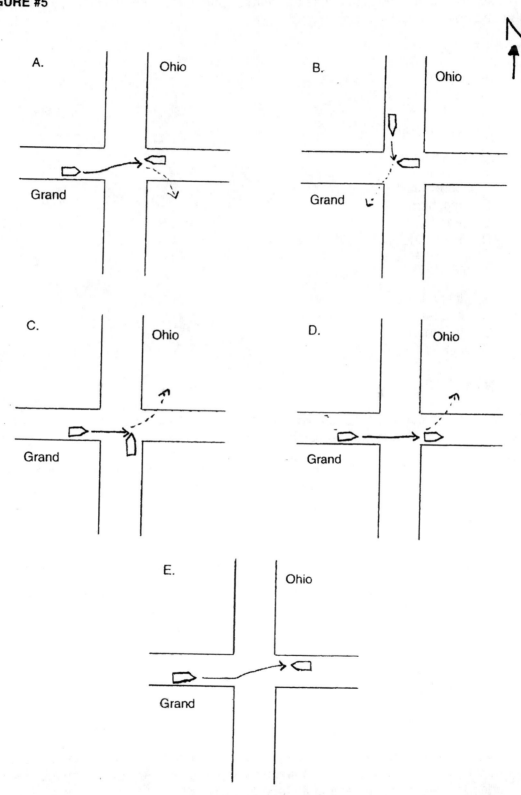

Questions 20-22

In questions 20 through 22, choose the word or phrase CLOSEST in meaning to the word or phrase printed in capital letters.

20. PETITION

    A. appeal
    B. law
    C. oath
    D. opposition

20._____

21. MALPRACTICE

    A. commission
    B. mayhem
    C. error
    D. misconduct

21._____

22. EXONERATE

    A. incriminate
    B. accuse
    C. lengthen
    D. acquit

22._____

Questions 23-25

Questions 23 through 25 measure your ability to do fieldwork-related arithmetic. Each question presents a separate arithmetic problem for you to solve.

23. Officers Lane and Bryant visited another city as part of an investigation. Because each is from a different precinct, they agree to split all expenses. With her credit card, Lane paid $70 for food and $150 for lodging. Bryant wrote checks for gas ($50) and entertainment ($40).
How much does Bryant owe Lane?

    A. $65   B. $90   C. $155   D. $210

23._____

24. In a remote mountain pass, two search-and-rescue teams, one from Silverton and one from Durango, combine to look for a family that disappeared in a recent snowstorm. The combined team is composed of 20 members. Which of the following statements could NOT be true?

    A. The Durango team has a dozen members
    B. The Silverton team has only one member
    C. The Durango team has two more members than the Silverton team
    D. The Silverton team has one more member than the Durango team

24._____

25. Three people in the department share a vehicle for a period of one year. The average number of miles traveled per month by each person is 150. How many miles will be added to the car's odometer at the end of the year?

    A. 1,800   B. 2,400   C. 3,600   D. 5,400

25._____

KEY (CORRECT ANSWERS)

1.	D		11.	D
2.	C		12.	C
3.	A		13.	D
4.	A		14.	D
5.	A		15.	B
6.	D		16.	E
7.	B		17.	A
8.	C		18.	C
9.	B		19.	D
10.	D		20.	A

21. D
22. D
23. A
24. D
25. D

———

TEST 2

DIRECTIONS: Each question or incomplete statement is followed by several suggested answers or completions. Select the one that BEST answers the question or completes the statement. *PRINT THE LETTER OF THE CORRECT ANSWER IN THE SPACE AT THE RIGHT.*

Questions 1-9

Questions 1 through 9 measure your ability to (1) determine whether statements from witnesses say essentially the same thing and (2) determine the evidence needed to make it reasonably certain that a particular conclusion is true.

To do well on this part of the test, you do NOT have to have a working knowledge of police procedures and techniques. Nor do you have to have any more familiarity with criminals and criminal behavior than that acquired from reading newspapers, listening to radio or watching TV. To do well in this part, you must read and reason carefully.

1. Which of the following pairs of statements say essentially the same thing in two different   1.____
    ways?

        I.   If there is life on Mars, we should fund NASA.
            Either there is life on Mars, or we should not fund NASA.
        II. All Eagle Scouts are teenage boys.
            All teenage boys are Eagle Scouts.

    A. I only
    B. I and II
    C. II only
    D. Neither I nor II

2. Which of the following pairs of statements say essentially the same thing in two different   2.____
    ways?

        I.   If that notebook is missing its front cover, it definitely belongs to Carter.
            Carter's notebook is the only one missing its front cover.
        II. If it's hot, the pool is open.
            The pool is open if it's hot.

    A. I only
    B. I and II
    C. II only
    D. Neither I nor II

3. Which of the following pairs of statements say essentially the same thing in two different   3.____
    ways?

        I.   Nobody who works at the mill is without benefits.
            Everyone who works at the mill has benefits.
        II. We will fund the program only if at least 100 people sign the petition.
            Either we will fund the program or at least 100 people will sign the petition.

    A. I only
    B. I and II
    C. II only
    D. Neither I nor II

4. Which of the following pairs of statements say essentially the same thing in two different ways?   4.____

      I.   If the new parts arrive, Mr. Luther's request has been answered.
          Mr. Luther requested new parts to arrive.
      II. The machine's test cycle will not run unless the operation cycle is not running.
          The machine's test cycle must be running in order for the operation cycle to run.

  A. I only
  B. I and II
  C. II only
  D. Neither I nor II

5. Summary of Evidence Collected to Date: 5.____

      I.   To become a member of the East Side Crips, a kid must be either "jumped in" or steal a squad car without getting caught.
      II. Sid, a kid on the East Side, was caught stealing a squad car.

Prematurely Drawn Conclusion: Sid did not become a member of the East Side Crips. Which of the following pieces of evidence, if any, would make it *reasonably certain* that the conclusion drawn is true?

  A. "Jumping in" is not allowed in prison
  B. Sid was not "jumped in"
  C. Sid's stealing the squad car had nothing to do with wanting to join the East Side Crips
  D. None of these

6. Summary of Evidence Collected to Date: 6.____

      I.   Jones, a Precinct 8 officer, has more arrests than Smith.
      II. Smith and Watson have exactly the same number of arrests.

Prematurely Drawn Conclusion: Watson is not a Precinct 8 officer.
Which of the following pieces of evidence, if any, would make it *reasonably certain* that the conclusion drawn is true?

  A. All the officers in Precinct 8 have more arrests than Watson
  B. All the officers in Precinct 8 have fewer arrests than Watson
  C. Watson has fewer arrests than Jones
  D. None of these

7. Summary of Evidence Collected to Date: 7.____

      I.   Twenty one-dollar bills are divided among Frances, Kerry and Brian.
      II. If Kerry gives her dollar bills to Frances, then Frances will have more money than Brian.

Prematurely Drawn Conclusion: Frances has twelve dollars.
Which of the following pieces of evidence, if any, would make it *reasonably certain* that the conclusion drawn is true?

  A. If Brian gives his dollars to Kerry, then Kerry will have more money than Frances
  B. Brian has two dollars
  C. If Kerry gives her dollars to Brian, Brian will still have less money than Frances
  D. None of these

8. <u>Summary of Evidence Collected to Date:</u>
         I.   The street sweepers will be here at noon today.
        II.   Residents on the west side of the street should move their cars before noon.
    <u>Prematurely Drawn Conclusion:</u> Today is Wednesday.
    Which of the following pieces of evidence, if any, would make it *reasonably certain* that the conclusion drawn is true?

    A.   The street sweepers never sweep the east side of the street on Wednesday
    B.   The street sweepers arrive at noon every other day
    C.   There is no parking allowed on the west side of the street on Wednesday
    D.   None of these

9. <u>Summary of Evidence Collected to Date:</u>
    The only time the warning light comes on is when there is a power surge.
    <u>Prematurely Drawn Conclusion:</u> The warning light does not come on if the air conditioner is not running.
    Which of the following pieces of evidence, if any, would make it *reasonably certain* that the conclusion drawn is true?

    A.   The air conditioner does not turn on if the warning light is on
    B.   Sometimes a power surge is caused by the dishwasher
    C.   There is only a power surge when the air conditioner turns on
    D.   None of these

8._____

9._____

Questions 10-14

Questions 10 through 14 refer to Map #6 and measure your ability to orient yourself within a given section of town, neighborhood or particular area. Each of the questions describes a starting point and a destination. Assume that you are driving a car in the area shown on the map accompanying the questions. Use the map as a basis for the shortest way to get from one point to another without breaking the law.

On the map, a street marked by arrows, or by arrows and the words "One Way," indicates one-way travel, and should be assumed to be one-way for the entire length, even when there are breaks or jogs in the street. EXCEPTION: A street that does not have the same name over the full length.

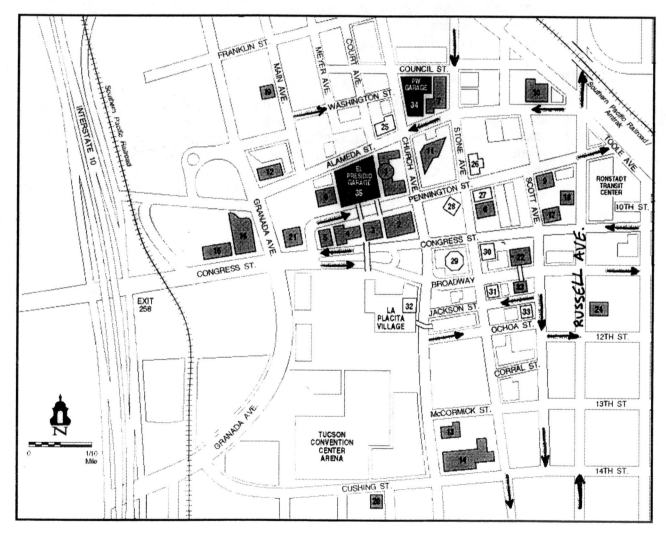

Map #6

PIMA COUNTY

1 Old Courthouse
2 Superior Court Building
3 Administration Building
4 Health and Welfare Building
5 Mechanical Building
6 Legal Services Building
7 County/City Public Works Center

CITY OF TUCSON

8 City Hall
9 City Hall Annex
10 Alameda Plaza City Court Building
11 Public Library - Main Branch
12 Tucson Water Building
13 Fire Department Headquarters
14 Police Department Building

10. The shortest legal way from the Public Library to the Alameda Plaza City Court Building is   10.____

    A.   north on Stone Ave., east on Alameda
    B.   south on Stone Ave., east on Congress, north on Russell Ave., west on Alameda
    C.   south on Stone Ave., east on Pennington, north on Russell Ave., west on Alameda
    D.   south on Church Ave., east on Pennington, north on Russell Ave., west on Alameda

11. The shortest legal way from City Hall to the Police Department is   11.____

    A.   east on Congress, south on Scott Ave., west on 14th
    B.   east on Pennington, south on Stone Ave.
    C.   east on Congress, south on Stone Ave.
    D.   east on Pennington, south on Church Ave.

12. The shortest legal way from the Tucson Water Building to the Legal Service Building is   12.____

    A.   south on Granada Ave., east on Congress, north to east on Pennington, south on Stone Ave.
    B.   east on Alameda, south on Church Ave., east on Pennington, south on Stone Ave.
    C.   north on Granada Ave., east on Washington, south on Church Ave., east on Pennington, south on Stone Ave.
    D.   south on Granada Ave., east on Cushing, north on Stone Ave.

13. The shortest legal way from the Tucson Convention Center Arena to the City Hall Annex is   13.____

    A.   west on Cushing, north on Granada Ave., east on Congress, east on Broadway, north on Scott Ave.
    B.   east on Cushing, north on Church Ave., east on Pennington
    C.   east on Cushing, north on Russell Ave., west on Pennington
    D.   east on Cushing, north on Stone Ave., east on Pennington

14. The shortest legal way from the Ronstadt Transit Center to the Fire Department is   14.____

    A.   west on Pennington , south on Stone Ave., west on McCormick
    B.   west on Congress, south on Russell Ave., west on 13th
    C.   west on Congress, south on Church Ave.
    D.   west on Pennington, south on Church Ave.

Questions 15-19

Questions 15 through 19 refer to Figure #6, on the following page, and measure your ability to understand written descriptions of events. Each question presents a description of an accident or event and asks you which of the five drawings in Figure #6 BEST represents it.

In the drawings, the following symbols are used:

Moving vehicle: ◊   Non-moving vehicle: ◆

Pedestrian or bicyclist: ●

The path and direction of travel of a vehicle or pedestrian is indicated by a solid line.

The path and direction of travel of each vehicle or pedestrian directly involved in a collision from the point of impact is indicated by a dotted line.

In the space at the right, print the letter of the drawing that best fits the descriptions written below:

15. A bicyclist heading southwest on Rose travels into the intersection, sideswipes a car that is heading east on Page, and veers right, leaving the roadway at the northwest corner of Page and Mill.

15.____

16. A driver traveling north on Mill swerves right to avoid a bicyclist that is traveling southwest on Rose. The driver strikes the rear end of a car parked on Rose. The bicyclist continues through the intersection and travels west on Page.

16.____

17. A bicyclist heading southwest on Rose travels into the intersection, sideswipes a car that is heading east on Page, and veers right, striking the rear end of a car parked in the westbound lane on Page.

17.____

18. A driver traveling east on Page swerves left to avoid a bicyclist that is traveling southwest on Rose. The driver strikes the rear end of a car parked on Mill. The bicyclist continues through the intersection and travels west on Page.

18.____

19. A bicyclist heading southwest on Rose enters the intersection and sideswipes a car that is swerving left to avoid her. The bicyclist veers left and collides with a car parked in the southbound lane on Mill. The driver of the car veers left and collides with a car parked in the northbound lane on Mill.

19.____

FIGURE #6

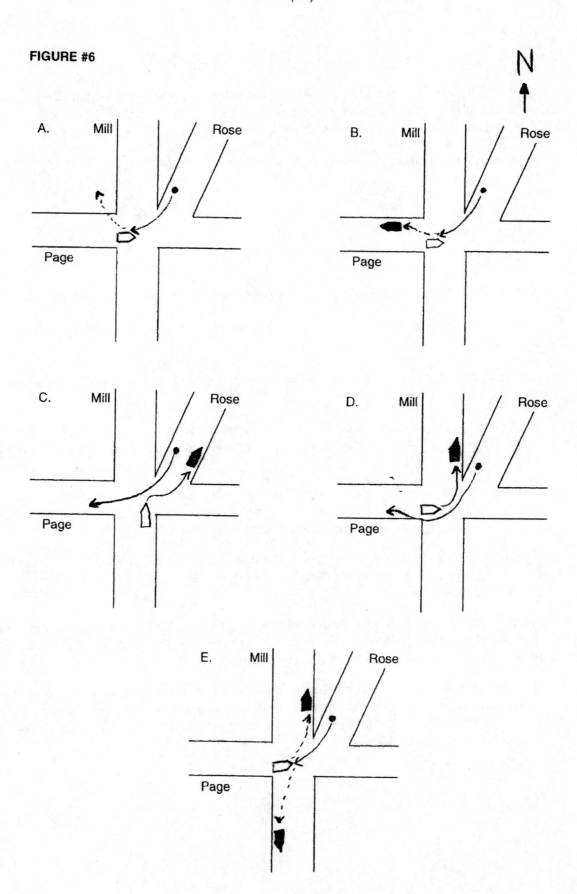

Questions 20-22

In questions 20 through 22, choose the word or phrase CLOSEST in meaning to the word or phrase printed in capital letters.

20. WAIVE

    A. cease
    B. surrender
    C. prevent
    D. die

20.____

21. DEPOSITION

    A. settlement
    B. deterioration
    C. testimony
    D. character

21.____

22. IMMUNITY

    A. exposure
    B. accusation
    C. protection
    D. exchange

22.____

Questions 23-25

Questions 23 through 25 measure your ability to do fieldwork-related arithmetic. Each question presents a separate arithmetic problem for you to solve.

23. Dean, a claims investigator, is reading a 445-page case record in his spare time at work. He has already read 157 pages. If Dean reads 24 pages a day, he should finish reading the rest of the record in _____ days.

    A. 7   B. 12   C. 19   D. 24

23.____

24. The Fire Department owns four cars. The Department of Sanitation owns twice as many cars as the Fire Department. The Department of Parks and Recreation owns one fewer car than the Department of Sanitation. The Department of Parks and Recreation is buying new tires for each of its cars. Each tire costs $100. How much is the Department of Parks and Recreation going to spend on tires?

    A. $400   B. $2,800   C. $3,200   D. $4,900

24.____

25. A dance hall is about 5,000 square feet. The local ordinance does not allow more than 50 people per every 100 square feet of commercial space. The maximum occupancy of the hall is

    A. 500   B. 2,500   C. 5,000   D. 25,000

25.____

KEY (CORRECT ANSWERS)

1.	D		11.	D
2.	B		12.	A
3.	A		13.	B
4.	A		14.	C
5.	B		15.	A
6.	D		16.	C
7.	D		17.	B
8.	D		18.	D
9.	C		19.	E
10.	C		20.	B

21.	C
22.	C
23.	B
24.	B
25.	B

PREPARING WRITTEN MATERIALS

EXAMINATION SECTION
TEST 1

DIRECTIONS: Each question consists of a sentence which may be classified appropriately under one of the following four categories:
   A.   Incorrect because of faulty grammar or sentence structure;
   B.   Incorrect because of faulty punctuation;
   C.   Incorrect because of faulty capitalization;
   D.   Correct.

Examine each sentence carefully. Then, in the space at the right, indicate the letter preceding the category which is the BEST of the four suggested above. Each incorrect sentence contains only one type of error. Consider a sentence correct if it contains no errors, although there may be other correct ways of expressing the same thought.

1.   All the employees, in this office, are over twenty-one years old.   1._____

2.   Neither the clerk nor the stenographer was able to explain what had happened.   2._____

3.   Mr. Johnson did not know who he would assign to type the order.   3._____

4.   Mr. Marshall called her to report for work on Saturday.   4._____

5.   He might of arrived on time if the train had not been delayed.   5._____

6.   Some employees on the other hand, are required to fill out these forms every month.   6._____

7.   The supervisor issued special instructions to his subordinates to prevent their making errors.   7._____

8.   Our supervisor Mr. Williams, expects to be promoted in about two weeks.   8._____

9.   We were informed that prof. Morgan would attend the conference.   9._____

10.   The clerks were assigned to the old building; the stenographers, to the new building.   10._____

11.   The supervisor asked Mr. Smith and I to complete the work as quickly as possible.   11._____

12.   He said, that before an employee can be permitted to leave, the report must be finished.   12._____

13.   An adding machine, in addition to the three typewriters, are needed in the new office.   13._____

14.   Having made many errors in her work, the supervisor asked the typist to be more careful.   14._____

15.   "If you are given an assignment," he said, "you should begin work on it as quickly as possible."   15._____

16.   All the clerks, including those who have been appointed recently are required to work on the new assignment.   16._____

17. The office manager asked each employee to work one Saturday a month. 17.____

18. Neither Mr. Smith nor Mr. Jones was able to finish his assignment on time. 18.____

19. The task of filing these cards is to be divided equally between you and he. 19.____

20. He is an employee whom we consider to be efficient. 20.____

21. I believe that the new employees are not as punctual as us. 21.____

22. The employees, working in this office, are to be congratulated for their work. 22.____

23. The delay in preparing the report was caused, in his opinion, by the lack of proper supervision and coordination. 23.____

24. John Jones accidentally pushed the wrong button and then all the lights went out. 24.____

25. The investigator ought to of had the witness sign the statement. 25.____

KEY (CORRECT ANSWERS)

1.	B		11.	A
2.	D		12.	B
3.	A		13.	A
4.	C		14.	A
5.	A		15.	D
6.	B		16.	B
7.	D		17.	C
8.	B		18.	D
9.	C		19.	A
10.	D		20.	D

21.	A
22.	B
23.	D
24.	D
25.	A

TEST 2

DIRECTIONS: Each of the following sentences may be classified under one of the following four options:
  A. Faulty; contains an error in grammar only
  B. Faulty; contains an error in spelling only
  C. Faulty; contains an error in grammar and an error in spelling
  D. Correct; contains no error in grammar or in spelling

Examine each sentence carefully to determine under which of the above four options it is BEST classified. Then, in the space at the right, write the letter preceding the option which is the best of the four listed above.

1. A recognized principle of good management is that an assignment should be given to whomever is best qualified to carry it out.   1._____

2. He considered it a privilege to be allowed to review and summarize the technical reports issued annually by your agency.   2._____

3. Because the warehouse was in an inaccessable location, deliveries of electric fixtures from the warehouse were made only in large lots.   3._____

4. Having requisitioned the office supplies, Miss Brown returned to her desk and resumed the computation of petty cash disbursements.   4._____

5. One of the advantages of this chemical solution is that records treated with it are not inflamable.   5._____

6. The complaint of this employee, in addition to the complaints of the other employees, were submitted to the grievance committee.   6._____

7. A study of the duties and responsibilities of each of the various categories of employees was conducted by an unprejudiced classification analyst.   7._____

8. Ties of friendship with this subordinate compels him to withold the censure that the subordinate deserves.   8._____

9. Neither of the agencies are affected by the decision to institute a program for rehabilitating physically handi-caped men and women.   9._____

10. The chairman stated that the argument between you and he was creating an intolerable situation.   10._____

Questions 11-25.

DIRECTIONS: Each of the following sentences may be classified under one of the following four options:

A. Correct
B. Sentence contains an error in spelling
C. Sentence contains an error in grammar
D. Sentence contains errors in both grammar and spelling.

11. He reported that he had had a really good time during his vacation although the farm was located in a very inaccessible portion of the country.   11._____

12. It looks to me like he has been fasinated by that beautiful painting.   12._____

13. We have permitted these kind of pencils to accumulate on our shelves, knowing we can sell them at a profit of five cents apiece any time we choose.   13._____

14. Believing that you will want an unexagerated estimate of the amount of business we can expect, I have made every effort to secure accurate figures.   14._____

15. Each and every man, woman and child in that untrameled wilderness carry guns for protection against the wild animals.   15._____

16. Although this process is different than the one to which he is accustomed, a good chemist will have no trouble.   16._____

17. Insensible to the fuming and fretting going on about him, the engineer continued to drive the mammoth dynamo to its utmost capacity.   17._____

18. Everyone had studied his lesson carefully and was consequently well prepared when the instructor began to discuss the fourth dimention.   18._____

19. I learned Johnny six new arithmetic problems this afternoon.   19._____

20. Athletics is urged by our most prominent citizens as the pursuit which will enable the younger generation to achieve that ideal of education, a sound mind in a sound body.   20._____

21. He did not see whoever was at the door very clearly but thinks it was the city tax appraiser.   21._____

22. He could not scarsely believe that his theories had been substantiated in this convincing fashion.   22._____

23. Although you have displayed great ingenuity in carrying out your assignments, the choice for the position still lies among Brown and Smith.   23._____

24. If they had have pleaded at the time that Smith was an accessory to the crime, it would have lessened the punishment.   24._____

25. It has proven indispensible in his compilation of the facts in the matter.   25._____

KEY (CORRECT ANSWERS)

1. A	11. A
2. D	12. D
3. B	13. C
4. D	14. B
5. B	15. D
6. A	16. C
7. D	17. A
8. C	18. B
9. C	19. C
10. A	20. A

21. B
22. D
23. C
24. D
25. B

TEST 3

DIRECTIONS: Questions 1 through 5 consist of sentences which may or may not contain errors in grammar or spelling or both. Sentences which do not contain errors in grammar or spelling or both are to be considered correct, even though there may be other correct ways of expressing the same thought. Examine each sentence carefully. Then, in the space at the right, write the letter of the answer which is the BEST of those suggested below:
    A. If the sentence is correct;
    B. If the sentence contains an error in spelling;
    C. If the sentence contains an error in grammar;
    D. If the sentence contains errors in both grammar and spelling.

1. Brown is doing fine although the work is irrevelant to his training.   1._____

2. The conference of sales managers voted to set its adjournment at one o'clock in order to give those present an opportunity to get rid of all merchandise.   2._____

3. He decided that in view of what had taken place at the hotel that he ought to stay and thank the benificent stranger who had rescued him from an embarassing situation.   3._____

4. Since you object to me criticizing your letter, I have no alternative but to consider you a mercenary scoundrel.   4._____

5. I rushed home ahead of schedule so that you will leave me go to the picnic with Mary.   5._____

Questions 6-15.

DIRECTIONS: Some of the following sentences contain an error in spelling, word usage, or sentence structure, or punctuation. Some sentences are correct as they stand although there may be other correct ways of expressing the same thought. All incorrect sentences contain only one error. Mark your answer to each question in the space at the right as follows:
    A. If the sentence has an error in spelling;
    B. If the sentence has an error in punctuation or capitalization;
    C. If the sentence has an error in word usage or sentence structure;
    D. If the sentence is correct.

6. Because the chairman failed to keep the participants from wandering off into irrelevant discussions, it was impossible to reach a consensus before the meeting was adjourned.   6._____

7. Certain employers have an unwritten rule that any applicant, who is over 55 years of age, is automatically excluded from consideration for any position whatsoever.   7._____

8. If the proposal to build schools in some new apartment buildings were to be accepted by the builders, one of the advantages that could be expected to result would be better communication between teachers and parents of schoolchildren.   8._____

9. In this instance, the manufacturer's violation of the law against deseptive packaging was discernible only to an experienced inspector.   9._____

10. The tenants' anger stemmed from the president's going to Washington to testify without consulting them first. 10._____

11. Did the president of this eminent banking company say; "We intend to hire and train a number of these disadvan-taged youths?" 11._____

12. In addition, today's confidential secretary must be knowledgable in many different areas: for example, she must know modern techniques for making travel arrangements for the executive. 12._____

13. To avoid further disruption of work in the offices, the protesters were forbidden from entering the building unless they had special passes. 13._____

14. A valuable secondary result of our training conferences is the opportunities afforded for management to observe the reactions of the participants. 14._____

15. Of the two proposals submitted by the committee, the first one is the best. 15._____

Questions 16-25.

DIRECTIONS: Each of the following sentences may be classified MOST appropriately under one of the following three categories:
A. Faulty because of incorrect grammar
B. Faulty because of incorrect punctuation
C. Correct

Examine each sentence. Then, print the capital letter preceding the BEST choice of the three suggested above. All incorrect sentences contain only one type of error. Consider a sentence correct if it contains none of the types of errors mentioned, even though there may be other ways of expressing the same thought.

16. He sent the notice to the clerk who you hired yesterday. 16._____

17. It must be admitted, however that you were not informed of this change. 17._____

18. Only the employees who have served in this grade for at least two years are eligible for promotion. 18._____

19. The work was divided equally between she and Mary. 19._____

20. He thought that you were not available at that time. 20._____

21. When the messenger returns; please give him this package. 21._____

22. The new secretary prepared, typed, addressed, and delivered, the notices. 22._____

23. Walking into the room, his desk can be seen at the rear. 23._____

24. Although John has worked here longer than she, he produces a smaller amount of work. 24._____

25. She said she could of typed this report yesterday. 25._____

KEY (CORRECT ANSWERS)

1.	D	11.	B
2.	A	12.	A
3.	D	13.	C
4.	C	14.	D
5.	C	15.	C
6.	A	16.	A
7.	B	17.	B
8.	D	18.	C
9.	A	19.	A
10.	D	20.	C

21. B
22. B
23. A
24. C
25. A

———

TEST 4

Questions 1-5.

DIRECTIONS: Each of the following sentences may be classified MOST appropriately under one of the following three categories:
   A. Faulty because of incorrect grammar
   B. Faulty because of incorrect punctuation
   C. Correct

Examine each sentence. Then, print the capital letter preceding the BEST choice of the three suggested above. All incorrect sentences contain only one type of error. Consider a sentence correct if it contains none of the types of errors mentioned, even though there may be other correct ways of expressing the same thought.

1. Neither one of these procedures are adequate for the efficient performance of this task.   1._____

2. The typewriter is the tool of the typist; the cash register, the tool of the cashier.   2._____

3. "The assignment must be completed as soon as possible" said the supervisor.   3._____

4. As you know, office handbooks are issued to all new employees.   4._____

5. Writing a speech is sometimes easier than to deliver it before an audience.   5._____

Questions 6-15.

DIRECTIONS: Each statement given in Questions 6 through 15 contains one of the faults of English usage listed below. For each, choose from the options listed the MAJOR fault contained.
   A. The statement is not a complete sentence.
   B. The statement contains a word or phrase that is redundant.
   C. The statement contains a long, less commonly used word when a shorter, more direct word would be acceptable.
   D. The statement contains a colloquial expression that normally is avoided in business writing.

6. The fact that this activity will afford an opportunity to meet your group.   6._____

7. Do you think that the two groups can join together for next month's meeting?   7._____

8. This is one of the most exciting new innovations to be introduced into our college.   8._____

9. We expect to consummate the agenda before the meeting ends tomorrow at noon.   9._____

10. While this seminar room is small in size, we think we can use it.   10._____

11. Do you think you can make a modification in the date of the Budget Committee meeting?   11._____

12. We are cognizant of the problem but we think we can ameliorate the situation.   12._____

13. Shall I call you around three on the day I arrive in the City?   13._____

14. Until such time that we know precisely that the students will be present. 14.____

15. The consensus of opinion of all the members present is reported in the minutes. 15.____

Questions 16-25.

DIRECTIONS: For each of Questions 16 through 25, select from the options given below the MOST applicable choice.
    A.   The sentence is correct.
    B.   The sentence contains a spelling error only.
    C.   The sentence contains an English grammar error only.
    D.   The sentence contains both a spelling error and an English grammar error.

16. Every person in the group is going to do his share. 16.____

17. The man who we selected is new to this University. 17.____

18. She is the older of the four secretaries on the two staffs that are to be combined. 18.____

19. The decision has to be made between him and I. 19.____

20. One of the volunteers are too young for this complecated task, don't you think? 20.____

21. I think your idea is splindid and it will improve this report considerably. 21.____

22. Do you think this is an exaggerated account of the behavior you and me observed this morning? 22.____

23. Our supervisor has a clear idea of excelence. 23.____

24. How many occurences were verified by the observers? 24.____

25. We must complete the typing of the draft of the questionaire by noon tomorrow. 25.____

KEY (CORRECT ANSWERS)

1.	A		11.	C
2.	C		12.	C
3.	B		13.	D
4.	C		14.	A
5.	A		15.	B
6.	A		16.	A
7.	B		17.	C
8.	B		18.	C
9.	C		19.	C
10.	B		20.	D

21.	B
22.	D
23.	B
24.	B
25.	B

___

PREPARING WRITTEN MATERIAL

PARAGRAPH REARRANGEMENT
COMMENTARY

The sentences which follow are in scrambled order. You are to rearrange them in proper order and indicate the letter choice containing the correct answer at the space at the right.

Each group of sentences in this section is actually a paragraph presented in scrambled order. Each sentence in the group has a place in that paragraph; no sentence is to be left out. You are to read each group of sentences and decide upon the best order in which to put the sentences so as to form as well-organized paragraph.

The questions in this section measure the ability to solve a problem when all the facts relevant to its solution are not given.

More specifically, certain positions of responsibility and authority require the employee to discover connections between events sometimes, apparently, unrelated. In order to do this, the employee will find it necessary to correctly infer that unspecified events have probably occurred or are likely to occur. This ability becomes especially important when action must be taken on incomplete information.

Accordingly, these questions require competitors to choose among several suggested alternatives, each of which presents a different sequential arrangement of the events. Competitors must choose the MOST logical of the suggested sequences.

In order to do so, they may be required to draw on general knowledge to infer missing concepts or events that are essential to sequencing the given events. Competitors should be careful to infer only what is essential to the sequence. The plausibility of the wrong alternatives will always require the inclusion of unlikely events or of additional chains of events which are NOT essential to sequencing the given events.

It's very important to remember that you are looking for the best of the four possible choices, and that the best choice of all may not even be one of the answers you're given to choose from.

There is no one right way to these problems. Many people have found it helpful to first write out the order of the sentences, as they would have arranged them, on their scrap paper before looking at the possible answers. If their optimum answer is there, this can save them some time. If it isn't, this method can still give insight into solving the problem. Others find it most helpful to just go through each of the possible choices, contrasting each as they go along. You should use whatever method feels comfortable, and works, for you.

While most of these types of questions are not that difficult, we've added a higher percentage of the difficult type, just to give you more practice. Usually there are only one or two questions on this section that contain such subtle distinctions that you're unable to answer confidently, and you then may find yourself stuck deciding between two possible choices, neither of which you're sure about.

———

Preparing Written Material

EXAMINATION SECTION
TEST 1

DIRECTIONS: The following groups of sentences need to be arranged in an order that makes sense. Select the letter preceding the sequence that represents the best sentence order. *PRINT THE LETTER OF THE CORRECT ANSWER IN THE SPACE AT THE RIGHT.*

Question 1   1._____

1. The ostrich egg shell's legendary toughness makes it an excellent substitute for certain types of dishes or dinnerware, and in parts of Africa ostrich shells are cut and decorated for use as containers for water.
2. Since prehistoric times, people have used the enormous egg of the ostrich as a part of their diet, a practice which has required much patience and hard work-to hard-boil an ostrich egg takes about four hours.
3. Opening the egg's shell, which is rock hard and nearly an inch thick, requires heavy tools, such as a saw or chisel; from inside, a baby ostrich must use a hornlike projection on its beak as a miniature pick-axe to escape from the egg.
4. The offspring of all higher-order animals originate from single egg cells that are carried by mothers, and most of these eggs are relatively small, often microscopic.
5. The egg of the African ostrich, however, weighs a massive thirty pounds, making it the largest single cell on earth, and a common object of human curiosity and wonder.

The best order is

A. 5 4 1 2 3
B. 1 4 5 3 2
C. 4 2 3 5 1
D. 4 5 2 3 1

Question 2   2._____

1. Typically only a few feet high on the open sea, individual tsunami have been known to circle the entire globe two or three times if their progress is not interrupted, but are not usually dangerous until they approach the shallow water that surrounds land masses.
2. Some of the most terrifying and damaging hazards caused by earthquakes are tsunami, which were once called "tidal waves"— a poorly chosen name, since these waves have nothing to do with tides.
3. Then a wave, slowed by the sudden drag on the lower part of its moving water column, will pile upon itself, sometimes reaching a height of over 100 feet.
4. Tsunami (Japanese for "great harbor wave") are seismic waves that are caused by earthquakes near oceanic trenches, and once triggered, can travel up to 600 miles an hour on the open ocean.
5. A land-shoaling tsunami is capable of extraordinary destruction; some tsunami have deposited large boats miles inland, washed out two-foot-thick seawalls, and scattered locomotive trains over long distances.

The best order is

A. 4 1 3 2 5
B. 1 3 4 2 5
C. 5 1 3 2 4
D. 2 4 1 3 5

Question 3   3.____

1. Soon, by the 1940's, jazz was the most popular type of music among American intellectu-
    als and college students.
2. In the early days of jazz, it was considered "lowdown" music, or music that was played only
    in rough, disreputable bars and taverns.
3. However, jazz didn't take long to develop from early ragtime melodies into more complex,
    sophisticated forms, such as Charlie Parker's "bebop" style of jazz.
4. After charismatic band leaders such as Duke Ellington and Count Basic brought jazz to a
    larger audience, and jazz continued to evolve into more complicated forms, white audi-
    ences began to accept and even to enjoy the new American art form.
5. Many white Americans, who then dictated the tastes of society, were wary of music that
    was played almost exclusively in black clubs in the poorer sections of cities and towns.

The best order is

    A. 5 4 3 2 1
    B. 2 5 3 4 1
    C. 4 5 3 1 2
    D. 1 2 4 3 5

Question 4   4.____

1. Then, hanging in a windless place, the magnetized end of the needle would always point to
    the south.
2. The needle could then be balanced on the rim of a cup, or the edge of a fingernail, but this
    balancing act was hard to maintain, and the needle often fell off.
3. Other needles would point to the north, and it was important for any traveler finding his way
    with a compass to remember which kind of magnetized needle he was carrying.
4. To make some of the earliest compasses in recorded history, ancient Chinese "magicians"
    would rub a needle with a piece of magnetized iron called a lodestone.
5. A more effective method of keeping the needle free to swing with its magnetic pull was to
    attach a strand of silk to the center of the needle with a tiny piece of wax.

The best order is

    A. 4 2 5 1 3
    B. 4 3 5 2 1
    C. 4 5 2 1 3
    D. 4 1 3 5 2

Question 5   5.____

1. The now-famous first mate of the *HMS Bounty*, Fletcher Christian, founded one of the
    world's most peculiar civilizations in 1790.
2. The men knew they had just committed a crime for which they could be hanged, so they set
    sail for Pitcairn, a remote, abandoned island in the far eastern region of the Polynesian
    archipelago, accompanied by twelve Polynesian women and six men.
3. In a mutiny that has become legendary, Christian and the others forced Captain Bligh into a
    lifeboat and set him adrift off the coast of Tonga in April of 1789.
4. In early 1790, the *Bounty* landed at Pitcairn Island, where the men lived out the rest of their
    lives and founded an isolated community which to this day includes direct descendants of
    Christian and the other crewmen.
5. The *Bounty*, commanded by Captain William Bligh, was in the middle of a global voyage,
    and Christian and his shipmates had come to the conclusion that Bligh was a reckless
    madman who would lead them to their deaths unless they took the ship from him.

The best order is

A. 4 5 3 2 1
B. 1 3 5 2 4
C. 1 5 3 2 4
D. 3 1 5 4 2

Question 6   6.____

1. But once the vines had been led to make orchids, the flowers had to be carefully hand-pol-
    linated, because unpollinated orchids usually lasted less than a day, wilting and dropping
    off the vine before it had even become dark.
2. The Totonac farmers discovered that looping a vine back around once it reached a five-foot
    height on its host tree would cause the vine to flower.
3. Though they knew how to process the fruit pods and extract vanilla's flavoring agent, the
    Totonacs also knew that a wild vanilla vine did not produce abundant flowers or fruit.
4. Wild vines climbed along the trunks and canopies of trees, and this constant upward
    growth diverted most of the vine's energy to making leaves instead of the orchid flowers
    that, once pollinated, would produce the flavorful pods.
5. Hundreds of years before vanilla became a prized food flavoring in Europe and the West-
    ern World, the Totonac Indians of the Mexican Gulf Coast were skilled cultivators of the
    vanilla vine, whose fruit they literally worshipped as a goddess.

The best order is

A. 2 3 4 1 5
B. 2 4 3 1 5
C. 5 3 4 2 1
D. 3 4 1 2 5

Question 7   7.____

1. Once airborne, the spider is at the mercy of the air currents—usually the spider takes a brief journey, traveling close to the ground, but some have been found in air samples collected as high as 10,000 feet, or been reported landing on ships far out at sea.
2. Once a young spider has hatched, it must leave the environment into which it was born as quickly as possible, in order to avoid competing with its hundreds of brothers and sisters for food.
3. The silk rises into warm air currents, and as soon as the pull feels adequate the spider lets go and drifts up into the air, suspended from the silk strand in the same way that a person might parasail.
4. To help young spiders do this, many species have adapted a practice known as "aerial dispersal," or, in common speech, "ballooning."
5. A spider that wants to leave its surroundings quickly will climb to the top of a grass stem or twig, face into the wind, and aim its back end into the air, releasing a long stream of silk from the glands near the tip of its abdomen.

The best order is

A. 5 4 2 3 1
B. 5 2 4 1 3
C. 2 5 4 3 1
D. 2 4 5 3 1

Question 8   8.____

1. For about a year, Tycho worked at a castle in Prague with a scientist named Johannes Kepler, but their association was cut short by another argument that drove Kepler out of the castle, to later develop, on his own, the theory of planetary orbits.
2. Tycho found life without a nose embarrassing, so he made a new nose for himself out of silver, which reportedly remained glued to his face for the rest of his life.
3. Tycho Brahe, the 17th-century Danish astronomer, is today more famous for his odd and arrogant personality than for any contribution he has made to our knowledge of the stars and planets.
4. Early in his career, as a student at Rostock University, Tycho got into an argument with the another student about who was the better mathematician, and the two became so angry that the argument turned into a sword fight, during which Tycho's nose was sliced off.
5. Later in his life, Tycho's arrogance may have kept him from playing a part in one of the greatest astronomical discoveries in history: the elliptical orbits of the solar system's planets.

The best order is

A. 1 4 2 3 5
B. 4 2 3 5 1
C. 4 2 1 3 5
D. 3 4 2 5 1

Question 9   9.____

1. The processionaries are so used to this routine that if a person picks up the end of a silk line and brings it back to the origin—creating a closed circle—the caterpillars may travel around and around for days, sometimes starving ar freezing, without changing course.
2. Rather than relying on sight or sound, the other caterpillars, who are lined up end-to-end behind the leader, travel to and from their nests by walking on this silk line, and each will reinforce it by laying down its own marking line as it passes over.
3. In order to insure the safety of individuals, the processionary caterpillar nests in a tree with dozens of other caterpillars, and at night, when it is safest, they all leave together in search of food.
4. The processionary caterpillar of the European continent is a perfect illustration of how much some insect species rely on instinct in their daily routines.
5. As they leave their nests, the processionaries form a single-file line behind a leader who spins and lays out a silk line to mark the chosen path.

The best order is

A. 4 3 5 2 1
B. 3 5 4 2 1
C. 3 5 2 1 4
D. 4 5 3 1 2

Question 10   10.____

1. Often, the child is also given a handcrafted walker or push cart, to provide support for its first upright explorations.
2. In traditional Indian families, a child's first steps are celebrated as a ceremonial event, rooted in ancient myth.
3. These carts are often intricately designed to resemble the chariot of Krishna, an important figure in Indian mythology.
4. The sound of these anklet bells is intended to mimic the footsteps of the legendary child Rama, who is celebrated in devotional songs throughout India.
5. When the child's parents see that the child is ready to begin walking, they will fit it with specially designed ankle bracelets, adorned with gently ringing bells.

The best order is

A. 2 3 4 1 5
B. 2 5 3 1 4
C. 5 4 1 3 2
D. 5 3 2 1 4

Question 11   11._____

1. The settlers planted Osage orange all across Middle America, and today long lines and
    rectangles of Osage orange trees can still be seen on the prairies, running along the former
    boundaries of farms that no longer exist.
2. After trying sod walls and water-filled ditches with no success, American farmers began to
    look for a plant that was adaptable to prairie weather, and that could be trimmed into a
    hedge that was "pig-tight, horse-high, and bull-strong."
3. The tree, so named because it bore a large (but inedible) fruit the size of an orange, was
    among the sturdiest and hardiest of American trees, and was prized among Native Ameri-
    cans for the strength and flexibility of bows which were made from its wood.
4. The first people to practice agriculture on the American flatlands were faced with an impor-
    tant problem: what would they use to fence their land in a place that was almost entirely
    without trees or rocks?
5. Finally, an Illinois farmer brought the settlers a tree that was native to the land between the
    Red and Arkansas rivers, a tree called the Osage orange.

The best order is

A. 2 1 5 3 4
B. 1 2 3 4 5
C. 4 2 5 3 1
D. 4 2 1 3 5

Question 12   12._____

1. After about ten minutes of such spirited and complicated activity, the head dancer is free to
    make up his or her own movements while maintaining the interest of the New Year's crowd.
2. The dancer will then perform a series of leg kicks, while at the same time operating the
    lion's mouth with his own hand and moving the ears and eyes by means of a string which is
    attached to the dancer's own mouth.
3. The most difficult role of this dance belongs to the one who controls the lion's head; this
    person must lead all the other "parts" of the lion through the choreographed segments of
    the dance.
4. The head dancer begins with a complex series of steps, alternately stepping forward with
    the head raised, and then retreating a few steps while lowering the head, a movement that
    is intended to create the impression that the lion is keeping a watchful eye for anything evil.
5. When performing a traditional Chinese New Year's lion dance, several performers must fit
    themselves inside a large lion costume and work together to enact different parts of the
    dance.

The best order is

A. 5 3 4 2 1
B. 3 4 2 5 1
C. 3 1 5 4 2
D. 4 2 3 5 1

Question 13   13._____

1. For many years the shell of the chambered nautilus was treasured in Europe for its beauty
    and intricacy, but collectors were unaware that they were in possession of the structure that
    marked a "missing link" in the evolution of marine mollusks.
2. The nautilus, however, evolved a series of enclosed chambers in its shell, and invented a
    new use for the structure: the shell began to serve as a buoyancy device.
3. Equipped with this new flotation device, the nautilus did not need the single, muscular foot
    of its predecessors, but instead developed flaps, tentacles, and a gentle form of jet propul-
    sion that transformed it into the first mollusk able to take command of its own destiny and
    explore a three-dimensional world.
4. By pumping and adjusting air pressure into the chambers, the nautilus could spend the day
    resting on the bottom, and then rise toward the surface at night in search of food.
5. The nautilus shell looks like a large snail shell, similar to those of its ancestors, who used
    their shells as protective coverings while they were anchored to the sea floor.

The best order is

A. 5 2 4 1 3
B. 5 1 2 3 4
C. 1 2 5 3 4
D. 1 5 2 4 3

Question 14   14._____

1. While France and England battled for control of the region, the Acadiens prospered on the
    fertile farmland, which was finally secured by England in 1713.
2. Early in the 17th century, settlers from western France founded a colony called Acadie in
    what is now the Canadian province of Nova Scotia.
3. At this time, English officials feared the presence of spies among the Acadiens who might
    be loyal to their French homeland, and the Acadiens were deported to spots along the
    Atlantic and Caribbean shores of America.
4. The French settlers remained on this land, under English rule, for around forty years, until
    the beginning of the French and Indian War, another conflict between France and England.
5. As the Acadien refugees drifted toward a final home in southern Louisiana, neighbors
    shortened their name to "Cadien," and finally "Cajun," the name which the descendants of
    early Acadiens still call themselves.

The best order is

A. 1 4 2 3 5
B. 2 1 3 5 4
C. 2 1 4 3 5
D. 5 2 3 4 1

Question 15   15._____

1. Traditional households in the Eastern and Western regions of Africa serve two meals a day-one at around noon, and the other in the evening.
2. The starch is then used in the way that Americans might use a spoon, to scoop up a portion of the main dish on the person's plate.
3. The reason for the starch's inclusion in every meal has to do with taste as well as nutrition; African food can be very spicy, and the starch is known to cool the burning effect of the main dish.
4. When serving these meals, the main dish is usually served on individual plates, and the starch is served on a communal plate, from which diners break off a piece of bread or scoop rice or fufu in their fingers.
5. The typical meals usually consist of a thick stew or soup as the main course, and an accompanying starch—either bread, rice, or fufu, a starchy grain paste similar in consistency to mashed potatoes.

The best order is

     A.   5 2 3 4 1
     B.   5 1 4 3 2
     C.   1 4 5 3 2
     D.   1 5 4 2 3

Question 16   16._____

1. In the early days of the American Midwest, Indiana settlers sometimes came together to hold an event called an apple peeling, where neighboring settlers gathered at the homestead of a host family to help prepare the hosts' apple crop for cooking, canning, and making apple butter.
2. At the beginning of the event, each peeler sat down in front of a ten- or twenty-gallon stone jar and was given a crock of apples and a paring knife.
3. Once a peeler had finished with a crock, another was placed next to him; if the peeler was an unmarried man, he kept a strict count of the number of apples he had peeled, because the winner was allowed to kiss the girl of his choice.
4. The peeling usually ended by 9:30 in the evening, when the neighbors gathered in the host family's parlor for a dance social.
5. The apples were peeled, cored, and quartered, and then placed into the jar.

The best order is

     A.   1 5 3 4 2
     B.   2 5 3 4 1
     C.   1 2 5 3 4
     D.   2 1 5 4 3

Question 17   17._____

1. If your pet turtle is a land turtle and is native to temperate climates, it will stop eating some time in October, which should be your cue to prepare the turtle for hibernation.
2. The box should then be covered with a wire screen, which will protect the turtle from any rodents or predators that might want to take advantage of a motionless and helpless animal.
3. When your turtle hasn't eaten for a while and appears ready to hibernate, it should be moved to its winter quarters, most likely a cellar or garage, where the temperature should range between 40° and 45° F.
4. Instead of feeding the turtle, you should bathe it every day in warm water, to encourage the turtle to empty its intestines in preparation for its long winter sleep.
5. Here the turtle should be placed in a well-ventilated box whose bottom is covered with a moisture-absorbing layer of clay beads, and then filled three-fourths full with almost dry peat moss or wood chips, into which the turtle will burrow and sleep for several months.

The best order is

A. 1 4 3 5 2
B. 3 4 2 5 1
C. 3 2 4 1 5
D. 4 5 2 3 1

Question 18   18._____

1. Once he has reached the nest, the hunter uses two sturdy bamboo poles like huge chopsticks to pull the nest away from the mountainside, into a large basket that will be lowered to people waiting below.
2. The world's largest honeybees colonize the Nepalese mountainsides, building honeycombs as large as a person on sheer rock faces that are often hundreds of feet high.
3. In the remote mountain country of Nepal, a small band of "honey hunters" carry out a tradition so ancient that 10,000 year-old drawings of the practice have been found in the caves of Nepal.
4. To harvest the honey and beeswax from these combs, a honey hunter climbs above the nests, lowers a long bamboo-fiber ladder over the cliff, and then climbs down.
5. Throughout this dangerous practice, the hunter is stung repeatedly, and only the veterans, with skin that has been toughened over the years, are able to return from a hunt without the painful swelling caused by stings.

The best order is

A. 2 4 3 5 1
B. 2 4 1 5 3
C. 5 3 2 4 1
D. 3 2 4 1 5

Question 19   19.____

1. After the Romans left Britain, there were relentless attacks on the islands from the barbarian tribes of northern Germany–the Angles, Saxons, and Jutes.
2. As the empire weakened, Roman soldiers withdrew from Britain, leaving behind a country
   that continued to practice the Christian religion that had been introduced by the Romans.
3. Early Latin writings tell of a Christian warrior named Arturius (Arthur, in English) who led
   the British citizens to defeat these barbarian invaders, and brought an extended period of
   peace to the lands of Britain.
4. Long ago, the British Isles were part of the far-flung Roman Empire that extended across
   most of Europe and into Africa and Asia.
5. The romantic legend of King Arthur and his knights of the Round Table, one of the most
   popular and widespread stories of all time, appears to have some foundation in history.

The best order is

A. 5 4 3 2 1
B. 5 4 2 1 3
C. 4 5 2 3 1
D. 4 3 2 1 5

Question 20   20.____

1. The cylinder was allowed to cool until it sould stand on its own, and then it was cut from the
   tube and split down the side with a single straight cut.
2. Nineteenth-century glassmakers, who had not yet discovered the glazier's modern techniques for making panes of glass, had to create a method for converting their blown glass
   into flat sheets.
3. The bubble was then pierced at the end to make a hole that opened up while the glassmaker gently spun it, creating a cylinder of glass.
4. Turned on its side and laid on a conveyor belt, the cylinder was strengthened, or tempered,
   by being heated again and cooled very slowly, eventually flattening out into a single rectangular piece of glass.
5. To do this, the glassmaker dipped the end of a long tube into melted glass and blew into the
   other end of the tube, creating an expanding bubble of glass.

The best order is

A. 2 5 3 4 1
B. 2 4 5 3 1
C. 3 5 2 4 1
D. 3 1 4 5 2

Question 21   21._____

   1.   The splints are almost always hidden, but horses are occasionally born whose splinted toes project from the leg on either side, just above the hoof.

   2.   The second and fourth toes remained, but shrank to thin splints of bone that fused invisibly to the horse's leg bone.

   3.   Horses are unique among mammals, having evolved feet that each end in what is essentially a single toe, capped by a large, sturdy hoof.

   4.   Julius Caesar, an emperor of ancient Rome, was said to have owned one of these three-toed horses, and considered it so special that he would not permit anyone else to ride it.

   5.   Though the horse's earlier ancestors possessed the traditional mammalian set of five toes on each foot, the horse has retained only its third toe; its first and fifth toes disappeared completely as the horse evolved.

The best order is

   A.   3 5 2 1 4
   B.   5 3 2 4 1
   C.   3 2 5 1 4
   D.   5 2 3 1 4

Question 22   22._____

   1.   The new building materials—some of which are twenty feet long, and weigh nearly six tons—were transported to Pohnpei on rafts, and were brought into their present position by using hibiscus fiber ropes and leverage to move the stone columns upward along the inclined trunks of coconut palm trees.

   2.   The ancestors built great fires to heat the stone, and then poured cool seawater on the columns, which caused the stone to contract and split along natural fracture lines.

   3.   The now-abandoned enclave of Nan Madol, a group of 92 man-made islands off the shore of the Micronesian island of Pohnpei, is estimated to have been built around the year 500 A.D.

   4.   The islanders say their ancestors quarried stone columns from a nearby island, where large basalt columns were formed by the cooling of molten lava.

   5.   The structures of Nan Madol are remarkable for the sheer size of some of the stone "logs" or columns that were used to create the walls of the offshore community, and today anthropologists can only rely on the information of existing local people for clues about how Nan Madol was built.

The best order is

   A.   5 4 3 2 1
   B.   5 3 1 4 2
   C.   3 5 4 2 1
   D.   3 1 4 2 5

Question 23   23.____

1. One of the most easily manipulated substances on earth, glass can be made into ceramic tiles that are composed of over 90% air.
2. NASA's space shuttles are the first spacecraft ever designed to leave and re-enter the earth's atmosphere while remaining intact.
3. These ceramic tiles are such effective insulators that when a tile emerges from the oven in which it was fired, it can be held safely in a person's hand by the edges while its interior still glows at a temperature well over 2000° F.
4. Eventually, the engineers were led to a material that is as old as our most ancient civilizationsglass.
5. Because the temperature during atmospheric re-entry is so incredibly hot, it took NASA's engineers some time to find a substance capable of protecting the shuttles.

The best order is

A. 5 2 1 3 4
B. 2 5 4 1 3
C. 2 3 1 2 5
D. 5 4 3 1 2

Question 24   24.____

1. The secret to teaching any parakeet to talk is patience, and the understanding that when a bird "talks," it is simply imitating what it hears, rather than putting ideas into words.
2. You should stay just out of sight of the bird and repeat the phrase you want it to learn, for at least fifteen minutes every morning and evening.
3. It is important to leave the bird without any words of encouragement or farewell; otherwise it might combine stray remarks or phrases, such as "Good night," with the phrase you are trying to teach it.
4. For this reason, to train your bird to imitate your words you should keep it free of any distractions, especially other noises, while you are giving it "lessons."
5. After your repetition, you should quietly leave the bird alone for a while, to think over what it has just heard.

The best order is

A. 1 4 2 5 3
B. 1 2 4 3 5
C. 3 2 1 5 4
D. 3 1 5 4 2

Question 25

1. As a school approaches, fishermen from neighboring communities join their fishing boats together as a fleet, and string their gill nets together to make a huge fence that is held up by cork floats.
2. At a signal from the party leaders, or *nakura,* the family members pound the sides of the boats or beat the water with long poles, creating a sudden and deafening noise.
3. The fishermen work together to drag the trap into a half-circle that may reach 300 yards in diameter, and then the families move their boats to form the other half of the circle around the school of fish.
4. The school of fish flee from the commotion into the awaiting trap, where a final wall of net is thrown over the open end of the half-circle, securing the day's haul.
5. Indonesian people from the area around the Sulu islands live on the sea, in floating villages made of lashed-together or stilted homes, and make much of their living by fishing their home waters for migrating schools of snapper, scad, and other fish.

The best order is

A. 1 5 3 4 2
B. 1 2 4 3 5
C. 5 1 2 3 4
D. 5 1 3 2 4

_____

KEY (CORRECT ANSWERS)

1.	D	11.	C
2.	D	12.	A
3.	B	13.	D
4.	A	14.	C
5.	C	15.	D
6.	C	16.	C
7.	D	17.	A
8.	D	18.	D
9.	A	19.	B
10.	B	20.	A

21.	A
22.	C
23.	B
24.	A
25.	D

_____

Food Sanitation Guide

INTRODUCTION

Restaurants, hotel and catering services in the country and the city serve millions of meals daily. This places tremendous responsibility upon them in safeguarding public health by preparing and serving only wholesome foods.

There are a number of cardinal principles which must be observed in preparing and serving wholesome foods. The bacterial contamination of these foods can be kept at a minimum if these principles are followed.

The food-handler must always be aware that he may contaminate the product by poor personal hygiene and work habits. He must always keep his person clean and work tools in a clean and sanitary condition.

Food must be stored in such a manner as to protect it from contamination. Unfortunately unless the food is sterilized, which is rarely practical, the presence of some bacteria is unavoidable. In order to keep their growth to a minimum, proper time and temperature control methods must be practiced.

Special care must be taken in the handling of foods which are to be served without further heat treatment. Ready-to-eat foods must not be subjected to contamination by coming into contact with unprocessed or partially processed foodstuffs or unsanitized work surfaces and implements.

Wholesome foods cannot be prepared in a dirty plant. The importance of good housekeeping cannot be minimized as a factor in the production of wholesome foods.

These general principles are more fully developed in the guide that follows.

I. Food Storage

The recommendations and prohibitions made below, if followed, will result in a wholesome and bacterialogically sound food product.

A. Dry Storage Foods

1. Dry stored foods are to be protected against contamination by insects, rodents, dust and other types of dirt.
2. All food storage containers should be properly labeled.

B. Cold Storage

1. Frozen foods

(a) During storage frozen foods are to be completely frozen until ready for use. (0° F)

(b) The freezer should be equipped with a thermometer so freezer temperatures can be determined without entering the holding box.

(c) Foods are to be stored in an orderly manner to assure cold air circulation and are not to be stored directly on the floor.

C. Chilled Foods

1. Chilled foods should be kept at 45oF or less at all times. This may be done by the use of a walk-in refrigerator, reach-in refrigerator, refrigerated show cases, refrigerator counter and refrigerated tables, etc.
2. Refrigerators should be supplied with appropriate thermometers.
3. Containers holding foods should not be stored so that the bottom surface of the container rests on the surface of the food product in the container below it.
4. Cooked foods should be stored so that they do not become contaminated by raw foods.
5. All foodstuffs should be stored in such a manner as to protect them from contamination.

D. Storage of Hot Foods

Foods to be served hot soon after cooking should not at any time be allowed to drop below an internal temperature of 14° F. If food is not to be served immediately upon completion of the cooking, it may be kept at temperatures in excess of 14° F by the use of warming cabinets, steamtables, chafing dishes or any other devices suitable for these purposes. Hot perishable foods are not to be kept at room temperature when the internal and surface temperature of the food falls below 140° F. Rare roast beef can be an exception to this. (See handling of rare roast beef, Pages 7-9)

II. Cleaning and Sanitization of Equipment and Kitchen Utensils

Equipment, utensils and work surfaces which come in contact with food should be thoroughly cleaned and sanitized before and after food preparation.

A. Methods of Cleaning and Sanitizing

Prior to washing, manually remove all adhering food particles. Then wash, using a suitable soap or detergent, and hot water liberally applied by manual or mechanical means. After rinsing and removing all visible dirt and grease, sanitize using one of the following methods:

1. Heat Sanitization

(a) Clean hot water, 170° F or more, applied to all surfaces of the equipment or utensils for at least 30 seconds.

2. Chemical Sanitization

(a) Apply a commercial preparation (Sodium Hypochlorite type) being sure to follow label directions.

(b) If a commercial product is not available or desired, a suitable solution may be prepared by mixing 1/2 ounce of household bleach, (5.25% Sodium Hypochlorite) in one gallon of lukewarm water (do not use hot water). Flood the surfaces of the equipment and utensils with this solution for at least one minute. Do not rinse or wipe after this operation.

If necessary to dry, air dry. Do not use a solution which is more than two hours old. If more solution is required, prepare a fresh supply.

III. Principles of Food Preparation and Services

During food preparation, improper techniques may contaminate the product with disease-causing organisms. It is for this purpose that sanitary procedures must be observed. Listed below are some principles which should be followed.

A. Food that is to be served cold should be kept cold (45° F or less) through all stages of storage, processing, and serving. Thawing of frozen foods should be accomplished in such a manner so as to keep the surface and internal temperatures of the product 45° F or less at all times. If frozen food is to be thawed in water, running cold water is to be used.

B. Foods to be served hot are to be kept so that the internal and surface temperatures do not fall below 140° F. (See handling of rare roast beef - Pages 7-9). Care must be taken in the cooling of hot foods so they do not become contaminated by dust, contact with work clothes, human contact, etc. Cooling should be accomplished as quickly as possible by the use of fans, refrigeration, etc. To determine the temperature of foods, a food thermometer is to be used. (Hands are not to be used).

C. Partially processed and leftover foods are to be refrigerated at 45° F. or below. Just prior to service they are to be removed from the refrigerator and heated rapidly to serving temperatures so that the internal temperatures are not less than 140° F.

D. The holding of perishable foods between the temperatures of 140° and 45° F is to be kept at a minimum.

E. Contact of ready-to-eat foods with bare hands should be kept at an irreducible minimum and utensils should be used whenever possible.

F. Ready-to-eat foods should not be contaminated by coming in contact with work surfaces, equipment, utensils or hands previously in contact with raw foods until such surfaces, etc. have been cleaned and sanitized.

G. Do not place packing cases and cans on food work surfaces.

H. When necessary to taste foods during processing, a clean sanitized utensil should be used. When tasting again, either re-clean and re-sanitize utensil, or use another sanitized utensil.

I. Foods are to be cooked and processed as close to the time of service as possible.

J.   Menu planning should be such as to prevent excessive leftovers, and leftovers are not to be pooled with fresh foods during storage.

IV.   Transportation of Foods

In some food operations, it is necessary to transport food from a central kitchen (commissary) to an establishment where it is finally served. The food transported can be in a ready-to-eat state or a pre-cooked stage, which is finally processed at the place of service. The following practices should be observed to see that contamination is not introduced or possible previous bacterial contamination not afforded means for extensive multiplication during this period.

1. Transporting containers and vehicles should be clean and of sanitary design to facilitate cleaning.
2. Transporting containers and vehicles should have acceptable refrigerating and/or heating facilities for maintaining food at cold (45° F or below) or hot (above 140°) temperatures while in transit.
3. Food stored in transporting containers and vehicles should be protected from contamination.
4. A minimum amount of time is to be taken for the loading and unloading of foods from transporting vehicles so foods will not be exposed to adverse temperatures and conditions.

V.   Food Processing Techniques Relative to Specific Types of Service

    A.   Displayed Food (Buffet, Smorgasbord, etc.)

1. Hot foods are to be kept at or above 140° F on the display table by use of chafing dishes, steam tables or other suitable methods.
2. Cold foods are to be at temperatures 45° F. or less before being displayed and not to be exposed at room temperature for more than one hour unless some means is employed, (ice, mechanical refrigeration, etc.) to keep cold foods at or less than 45° F.
3. All foods displayed and, therefore, subject to contamination must be discarded at the conclusion of the buffet service.

    B.   Protein Type Salads (Tuna, Ham, Shrimp, Egg, Chicken, Lobster, etc.)

These salads are always served cold and, therefore, all salad ingredients except the seasoning and spices are to be chilled to 45° F or less before use. Celery, which is almost always a component of these salads, should be treated so as to minimize its bacterial content by the immersing of the chopped celery in boiling water, using a hand strainer or colander for 30 seconds and then chilling immediately by holding under running cold tap water.

Before the mixing operation, the previously washed can opener, and tops of cans and jars holding salad ingredients should be wiped with a clean cloth containing sanitizing solution. The salad ingredients should be mixed with clean, sanitizing equipment, (sanitary type masher, sanitary mixing bowl, stainless steel long handled spon or fork, mechanical tumbler type mixer, etc.). There should be an absolute minimum of bare hand contact with the equipment and ingredients. The mixing operation is to be completed as quickly as possible and the finished salad immediately served or refrigerated.

    C.   Additional Instructions Relative to Specific Salads

1. Shrimp and Lobster Salad
   Immerse shrimp, or lobster meat in boiling water for 30 seconds and then chill to 45° F or less before adding to salad. Fast chilling can be accomplished by placing the meat in shallow pans in the freezer or refrigerator or on top of cracked ice.

2. Egg Salad
   After removing shell, use a hand strainer or colander to immerse hard-boiled eggs in boiling water for 30 seconds and then chill to 45° F or less before adding to salad. Chill the eggs by refrigerating or by placing strainer containing them under running cold tap water.

3. Chicken and/or Turkey Salad
   After removal from bones, immerse chicken or turkey meat in boiling water or boiling stock for 30 seconds and then chill to 45° F. before adding to salad. Fast chilling can be accomplished by placing the meat in shallow pans in the freezer, refrigerator or on cracked ice.

4. Ham Salad
   Immerse diced ham in boiling water or boiling stock for 30 seconds and then chill to 45° F. or less before adding to salad. Fast chilling can be accomplished by the same method used for chicken and shrimp.

D. Hot Meats and Poultry Served from Steamtables or Other Suitable Warming Devices

1. Schedule the cooking of meats so they will be completed as close as possible to desired time of service.

2. Upon removal from the oven or stove, cooked meats are to be kept at an internal temperature of 140° F or higher in a steamtable or other suitable device.

3. Maintain the water in the steamtable at a temperature in excess of 180° F. The water must be brought to this temperature before any foods are placed therein. Water in the steamtable shall be kept at a steamtable depth so as to be in contact with the bottom and upper portions of the sides of the food container.

4. Refrigerated ready-to-eat cooked meats, especially leftovers, gravies and stocks, are to be heated rapidly to an internal temperatures of 165° F or higher before being placed in the steamtable or warming device. Hot stock or meat gravies may be used to reheat meats. Steamtables or other warming devices should never be used to heat up cold meats.

5. Cautions noted previously relative to hand contact, care of equipment storage, and menu planning should also be followed.

E. Roast Beef

   Because of consumer preference, roast beef is often served at an internal temperature of less than 140° F. Continuous warming and heating of this product, as for example on a steamtable, may not be practical as it causes the meat to become well done and thus less desirable to some consumers. It is, therefore, realized that instructions relative to mainte-nance of interior temperatures of meat cannot always be applied to this product. It is essential, therefore, that Roast Beef be cooked as close to time of service as possible. Great care must be taken to prevent contamination. At large banquets this roast is sometimes stored or "rested" for excessive lengths of time, during which bacterial growth can occur.

1. Bone in Standing Rib Roast

There are a number of methods to be used in the processing of this type of roast beef, which will help minimize bacterial contamination and growth.

(a) Method No. 1

The roast is boned and trimmed prior to cooking. Slicing is accomplished after cooking and immediately prior to serving. After removal from the oven, the surface temperature should be in excess of 140° F. This method minimizes the amount of handling after the cooking operations.

(b) Method No. 2

After cooking and storage the roast is boned, trimmed (all surfaces) and sliced immediately prior to service. This method removes almost all surface contamination.

(c) Method No. 3

The surface of the raw roast beef is coated with a concentration of coarse salt. The beef is cooked and stored with this coating intact and it is not removed until just prior to service, at which time, boning and trimming and slicing takes place. The salinity on the surface of the meat inhibits the bacterial growth. It has been found that after removal of the salt coating platibility of the meat is not impaired as there is practically no penetration of the salt into the edible portion of the meat.

2. Boneless Tied Roast Beef

This type of roast beef is commonly machine sliced and used on sandwiches and platters. As stated above this type of roast beef is often desired rate where high internal temperatures cannot be applied.

Menu planning should be such that the roast beef should be removed from the oven as close to the service time as possible.

After removing a large roast beef from the oven, it should be cut into smaller pieces, each not to exceed 6 pounds. The surface temperature of the meat should not fall below 140°., at which time the roast can be sliced for immediate service and placed in the refrigerator, warming oven or steamtable. The refrigerator temperature should be below 45° F and steamtable temperature in excess of 140° F.

It is suggested that only one piece of roast be kept for immediate service and the other pieces be stored in the refrigerator or warming device. At the end of the day any piece of roast beef which has been partially used should be considered as a leftover. This piece of meat must be refrigerated overnight, and before being reused it is to be heated to an internal temperature in excess of 165° F. It is realized that after cooking at these temperatures, this product cannot be served again as rare roast beef.

The slicing machine used for this product should be disassembled and cleaned at the end of the day's work, and left disassembled. Before beginning slicing operations the next day, it is to be sanitized and reassembled.

3. Steamship (Steamer) Beef Roasts

This type of roast consists of the whole beef round (top and bottom) usually served rare and stored at inadequate temperatures (less that 140°). This product is almost always hand carved. (The term hand carved is used to denote that it is not machine sliced). There is no need for hand contact inasmuch as this meat is sliced with the use of a chef's knife and fork and transferred to the sandwich or platter using these utensils. Since the normal means to prevent contamination cannot be excercised, it is mandatory that only properly sanitized equipment be used and the food-handler exert particular care not to contaminate the product. As stated above in paragraph 2, any unused portion of this roast should be refrigerated, and before being served again cooked to an internal temperature of 165° F. It is again realized that after recooking at this temperature this product cannot be served as rare roast beef.

F. Rare Steaks

If these are not cooked immediately prior to service, it is sometimes the practice to singe the outer surface of the meat, and then store it at room temperature until the time of service. It is cooked by broiling and served immediately.

For this type of meat service, it is important that the storage period is not over one hour, the meat does not come in contact with contaminated work surfaces or hands and the meat is subjected to sufficient surface terminal heat treatment just before serving.

G. Pre-cooked Hamburger Patties

It has become a practice in some restaurants to pre-cook hamburger patties, and store them in a warmer or above the stove or grill until needed for service. In most cases the temperatures and lengths of time the meat is kept can be such as to allow the growth of pathogenic organisms.

If this type of food preparation is practiced, extreme care should be taken to see that this product is not stored for more than one hour, the food is not contaminated by unclean hands or work surfaces, and it receives a thorough heat treatment (exceeding 16° F) just prior to its consumption.

H. Pre-cooked Chicken - (Barbecued Style)

This product, a whole eviscerated chicken of 2-3 pounds, is usually cooked in a rotisserie-type radiant heating device and stored for varying lengths of time and temperature. Again this type of food storage is advantageous for the growth of food poisoning organisms.

Precautions to be followed with this product are: all parts of the poultry are to be thoroughly and completely cooked (over 165° F); it is to be handled and stored so that it will not come in contact with contaminated hands or work surfaces; and it shall not be kept at temperatures between 45° F and 14° F for more than one hour anytime prior to consumption.

I.   Poultry Stuffing

Often times adequate internal temperatures are not obtained in the cooking of stuffed poultry. The temperature of the stuffing is such as to incubate rather than destroy bacteria. It is therefore advisable to cook the stuffing separately from the poultry. When this is done adequate temperatures (165° F) are reached in both the stuffing and poultry. Thereafter the stuffing should be handled and/ or stored in a manner similar to that noted previously for perishable protein foods.

J.   Custard-Filled Baked Goods

The problems with custard fillings arise after completion of the cooking operation during the cooling and handling period. The following recommendations are made:

1.   Utensils and receptacles must be sanitized as previously noted.
2.   The finished custard should be transferred to shallow stainless steel or aluminum trays to facilitate rapid cooling. It is important at this point not to contaminate the product with the foodhandler's hands or clothing.
3.   A long-bladed flexible spatula of sanitary construction should be used to scrape the residue from the cooking receptacle.
4.   The finished product should be refrigerated as quickly as possible and at no time should the product be exposed to room temperature for more than one hour.
5.   The shallow pans of custard should be covered with wax or other clean paper while cooling and while being stored in the refrigerator.
6.   Jelly-filling machines of sanitary design should be used. Multiple use pastry bags, after washing, are to be boiled or sanitized before use. A single service pastry bag can be fashioned out of wax or parchment paper. A desire method of filling eclair shells, cream puffs and similar type products is to cut the shell in half and apply the filling with a properly sanitized stainless steel spatula. This is the only method to be used in the production of napoleans.
7.   Butter cream which is to be used as an ingredient of custard should be handled with the same precautions as actual custard.
8.   The finished product, immediately after completion, must be kept under refrigeration (45° F or less) at all times until consumed. Commercial fillings, bavarian creams, etc., are often used instead of true custard. They are used, as per label directions, and are sometimes used with the addition of eggs, cream, butter cream, etc., depending on the recipe of the individual food processor. The same care, relative to the boiling and refrigeration of all ingredients, should be taken in the manufacture of these products as is observed with true custard.

K.   Deviled Eggs

It has been the experience of the Food Processing Control Unit that this product is needlessly contaminated by poor handling techniques. The following is suggested to minimize contamination.

1. This product is to be prepared as close to service as possible.
2. After the shell is removed from the egg, the peeled egg is to be placed in a strainer or colander and then in boiling water for not less than 30 seconds and then immediately plgced in running cold water and chilled to 45° F or less.
3. At this point, when it is unavoidable that the bare hands be used, it is mandatory that the food handler wash his hands thoroughly with a germicidal soap before proceeding with the process.
4. Whenever possible remove the yolk of the egg with a sanitized utensil, and when the yolk is mashed and mixed with seasonings, a sanitized utensi' is also to be used.
5. In extruding the mashed yolk, a single service pastry bag is recommended. If a multiple use bag is desired it is to be sanitized by heat or chemical treatment prior to use.
6. If the finished product is not used immediately, it is to be refrigerated at 45° F or less until served.

L. Fresh Pork Products

Though it has been previously mentioned that meats are to be cooked to proper internal temperatures, it is felt that an additional warning be given concerning fresh pork products. Government inspection of fresh pork is not a guarantee against trichina contamination of this product. The trichina are not readily detectable except by microscopic examination and then only if an infested area is examined. It is therefore mandatory that fresh pork products be cooked to an internal temperature of at least 150° F.

M. Chopped Liver

This perishable, popular product is ordinarily literally manhandled in processing. Inasmuch as most of the handling takes place after cooking and the product is served without further heat treatment extra precautions must be taken to minimize hand contact.

Equipment must be cleaned and sanitized before use. It is best to clean, sanitize, and assemble equipment immediately prior to use. Ingredients should not be touched with bare hands after cooking. Cooked liver is to be handled with sanitized equipment only. Hard boiled eggs, after shells are removed, are to be placed in a colander or strainer and immersed in boiling water for 30 seconds and then placed in running cold water and chilled to 45°F or less. The peeled eggs are then to be handled by implements only. After mixing, the finished chopped liver is to be placed in stainless steel serving containers or molds without use of bare hands. If hand molding is required for a decorative display this is to be done immediately prior to service.

VI. Plant Sanitation and Maintenance

The unclean and defective condition of the physical plant, walls, floors, ceilings, doors, windows, etc., can adversely affect the final product from a bacterial standpoint. Care should be taken to see that they are clean and maintained in such a manner as to facilitate proper plant sanitation. It is known that bacterial organisms will establish themselves on encrusted foods such as is found on walls, light switches, room and refrigerator door handles, and other surfaces touched by food-handlers. Improper wall, window and door maintenance, ineffective cleaning and poor garbage disposal methods can also lead to insect and rodent infestations. These well known vectors of disease organisms can introduce food poisoning bacteria to foodstuffs in the establishment. (When necessary acceptable insecticides and rodenticides can be used to prevent or exterminate an infestation of these pests. Care should be observed to see they do not come in contact with foodstuffs.)

Adequate amounts of hot and cold running water should be supplied at properly maintained fixtures, strategically placed in parts of the plant, i.e., toilets, food processing areas, utensil cleaning areas, etc. Such fixtures should also be supplied with detergents, bactericides, and single service hand towels.

Equipment should be of sanitary design and maintained in a sanitary condition, cleaned after use and sanitized before use. Open seams and worn or defective surfaces which allow food particles to accumulate and prevent proper cleaning should be repaired forthwith.

Self-inspection and cleaning schedules should be devised for all areas of the plant and equipment. At routine periods all areas should be inspected to detail and findings noted on a form devised for this purpose. Follow-up on findings should be made as soon as feasible.

Cleaning and maintenance should follow every major production period. If production is continuous for a 12 or 18 hour period, "down" periods should be incorporated in the work schedule to allow for this sanitation program.

VII.   Non-Commercial Food Operations

This guide is primarily for the use of the sanitarian and the operators of commercial food-processing establishments. In a city of this size, many large meals and buffets are prepared and served by private and volunteer organizations. These include church and synogogue socials, local charity and fund raising affairs, fraternal organizations, etc.

These types of affairs often lead to food-borne illnesses when proper precautions are not taken. It is therefore important that the recommendations contained herein also be practiced by these large non-commercial feeding operations.

VIII.   Assistance to Food Processors

Commercial and non-commercial food operations are urged to use the expertise of this department by calling upon us to discuss and analyze problems occuring in their food handling programs.

SALAD PREPARATION GUIDE

1. Refrigerate all salad ingredients except seasoning and spices overnight or chill to 45° F or lower before use.
2. Purchase a sanitizing solution or prepare one by mixing one or two ounces of bleach to a gallon of cold water. This solution is effective for approximately two hours. Prepare a fresh solution if further sanitization is needed.
3. Clean work surfaces, equipment and utensils (pots, pans, spoons, spatulas, etc.) with soap and hot water, rinse with clean water, and then give a final rinse with sanitizing solution. Stainless steel utensils and equipment are preferred in preparation of these foods.
4. Clean hands, fingernails, and arms thoroughly with ger-micidal soap and hot water and dry with single use paper towels.
5. Individuals preparing salads are not to perform other tasks while engaged in salad preparation.
6. Clean and sanitize tops of cans and jars before openings. Do not use fingers to pry off can lids or drain off liquid contents.
7. Place diced celery, including pre-cut packaged celery in a strainer and immerse in boiling water for 30 seconds; then chill to 45° F or less.
8. Use clean sanitized utensils in mixing and handling of foods. Avoid hand contact with foods.
9. Refrigerate final salad product immediately in shallow pans.
10. Salads placed in bain-marie cold plates should have a minimum internal temperature of 45° F.
11. Do not fill trays above spill line.

Control of Rodents and Insects
CONTROL OF RODENT INFESTATION AND HORBORAGE

In combating rodent infestation the use of cats, traps and poisons are only temporary expendients and do not eliminate rodent life completely from your premises. The best method of permanently eliminating them is to *build the out.* Rodent life exists in buildings because of favorable conditions that permit them to hide, nest and breed. They will not remain where safe shelter or food is not available. To combat infestation in your premises, it is necessary to be able to recognize rodent harborages or hiding places, both actual and potential as they are the conditions favoring rodent life and propagation. There are three general types of rodent harborages:

        1- Temporary
        2- Incidental
        3- Structural

TEMPORARY RODENT HARBORAGES

These are conditions arising out of failure to maintain premises in a clean and sanitary condition, or faulty methods of operation, housekeeping, or storage of stock.

EXAMPLES

I.   Mass storage of office supplies and old records, materials for repairs, food products or other store merchandise; boxes, crates, or cartons that are left undisturbed for periods of time and not rotated in use (using up older stock first).

II.   Unused or obsolete fixtures or equipment, especially those having drawers, compartments or other hollow enclosures.

III.   Miscellaneous junk, trash, odds and ends placed in closets, cellars, boiler rooms or out-of-the-way places, or portions of premises not in daily use having very little or no light.

IV.   Garbage cans left uncovered overnight or having poorly fitting covers, or in a defective leaking condition.

V.   Passageways used in transporting or storing garbage cans for removal, with spilled particles of food on floors, especially in corners.

VI.   Accumulations of rubbish at botton of airshafts, dumbwaiter or elevator shaft pits, under sidewalk or cellar window gratings, or other parts of premises not cleaned regularly.

PREVENTION

I.   Unused materials should be stored neatly and away from walls, allowing enough space for a man to pass around in cleaning and should preferably be stored sufficiently high above the floor to permit cleaning underneath. The amount stored should be minimized as much as possible, and it should be disturbed or its position changed at least every three weeks to prevent nesting of rodents.

II.   Avoid mass storage by arranging in rows with 2' wide aisles. If stock is placed on shelves, raise the lowest shelf about 6" to 8" above the floor. Remove all rubbish that usually accumulates about unused materials. Promptly clean up food scrapes that spill from garbage cans, or fall under, or behind slop sinks, equipment, and stock bins. (Rodents feed more readily on these than on bagged or packaged food supplies). Store all garbage in non-leaking metal receptacles with tightfitting covers.

III. Place soiled linen into suitable containers.

IV. Maintain clean and sanitary conditions at all times.

INCIDENTAL RODENT HARBORAGES

These are conditions arising from installing of fixtures or equipment incidental to their use on the premises, in such a manner that hollow spaces, enclosures, and inaccessible places are formed.

EXAMPLES

I. Fixtures, refrigerators, ovens, etc. not installed flush against walls but leaving a small space that is too narrow for proper inspection and cleaning.

II. Narrow spaces left between bottoms of counters, back bars, or other fixtures or equipment, and the floor.

III. Small spaces existing between ceilings and tops of fixtures, clothes lockers, refrigerators, closets and cabinets, large overhead pipes and ventilating ducts suspended a few inches from ceiling.

IV. Hollow partitions (double wall space).
Hollow furniture of fixtures with inaccessible enclosures. Boxed-in casings or sheathing around pillars, pipes, radiators, etc., forming hollow enclosures.

V. Bottom shelves, stock platforms or skids that are not set directly on the floor but allow a space of a few inches to exist underneath.

VI. Defective insulated sections of large refrigerators or pipe coverings (hollow enclosed spaces formed by damage to cork or asbestos).

VII. Loose foods stored in low, thin, wooden food bins, boxes, cartons, burlap bags, etc.

VIII. Partially enclosed spaces behind open metal grilles used on housing of motors or other mechanical equipment.

PREVENTION

I. Eliminate narrow, inaccessible spaces behind fixtures or equipment by placing flush against wall or leaving a space wide enough for inspection and cleaning. Solidly block out narrow spaces underneath, or install flush on floors or raise high enough for cleaning.

II. Avoid providing undisturbed rat runways in narrow space between ducts or long hoods, and the ceiling. Ducts should be placed flush against ceilings and preferably be found in shape, instead of square.

III. Remove decorative boxing-in around radiators, columns, etc., to avoid hollow enclosures, or protect gnawing margins with metal flashing extending at least 6" above the floors, if they must be sheathed for appearance, used sheet metal.

IV. Repair and securely close all breaks in insulation around pipes, refrigerators or cooling cabinets.

V. Line interiors of wooden bins with sheet metal, or store foods in rodent proof screening (mesh openings not greater than 1/4").

VI. Eliminate hollow spaces formed by false bottoms in counters, lockers, cabinets, back bars, etc.

VII. Alter hollow fixtures so that enclosures are exposed for easy cleaning.

STRUCTURAL RODENT HARBORAGES

These conditions are due to design or construction of a building that are defective from a rat-proof standpoint, or that developed during occupancy from failure to make proper repairs or to use rat-proof materials.

EXAMPLES

I. Openings made in outside building walls, around beams, or in interior walls, floors or ceiling for installation of pipes, cables, or conduits. They are made by plumbers, electricians, or other workmen. The openings are usually larger than necessary and the unused portions of holes are not closed up. Holes, large cracks, loose bricks, or other openings in floors, walls or ceilings are other examples.

II. Hollow spaces in double walls, between floor and ceiling of lower story, and in double ceilings of cellars.

III. Enclosed hollow spaces formed by sheathing the undersides of stairways, by installation of false floors in toilets, or by raised wooden floors over earthen floors of cellars.

IV. Entrance and cellar doors that are not tight-fitting or not provided with a proper door sill or saddle, permitting openings over 1/4" to exist and not protected around gnawing edges with metal flashing at least 6" above floor level.

V. Openings around ceiling or floor beams, or risers, where they pass through partitions.

VI. Openings of fans, ventilators, and louvers on the outside of buildings, or fancy metal grilles with openings over 1/4", not protected by rodent proof screening.

VII. Floor drain and sewer trap pits not kept clean and not provided with solid metal covers with preforations not exceeding 1/4". Cellar floors of earth, enabling rodents to burrow underneath.

PREVENTION

I. Promptly seal up all holes or openings around pipe lines or cables where they enter the building, with concrete mortar or cement mortar to which ground glass may be added for better results.

II. Place tight-fitting metal collars or flanges around pipes and risers. Provide escutcheon plates for all risers where they pass through floor slabs, unless same are waterproofed by pockets of mastic.

III. Seal up all openings around beams.

IV. Avoid using double-wall type construction with hollow interior spaces, or hollow tile block, hollow cement block, or similar material for partitions or walls of storage compartments or in cellars.

V. Inspect all parts of premises for holes and seal every opening in walls and ceilings with cement plastered smooth. Move away fixtures and stock that may hide holes in floors and use a flashlight so as not to miss any. Look for loose bricks, cracks or other openings in cellar foundation walls. Find all openings before rodents do. Inspect regularly and repair weak spots before actual breaks occur.

VI. Block out hollow spaces under raised wooden floors with concrete Refrigerators, ranges, ovens, etc., should be solidly based on concrete. Protect entrance, cellar doors, and windows with metal flashing around gnawing edges, and maintain in good repair.

VII. Replace earthen cellar floors with a floor of concrete at least 3" - 4" thick and tied securely into foundation walls.

VIII. Securely anchor window and door screens to the frames.

RODENT INFESTATION SURVEYS

In addition to trapping, surveys by operators will indicate presence and approximate extent of infestation, of which the following are some signs.

Excreta of Pellets

Physical state will indicate recent or old infestation. Soft, moist droppings indicate live rats or mice present, while hard and dry ones indicate old. Amount of droppings indicate heavy or light infestation. The size of pellets will show if the rodents are large or small; and if different sizes are present, it indicates litters of young are being reared.

Gnawings

If recent, will show fresh appearance of gnawings, shavings, debris or marks on food bags or containers, or damage to other merchandise, supplies or fixtures.

Rat Run

Difficult to tell by appearance if new or old. Use white chalk or paint on suspected rat run. The rat is a creature of habit and will continue to use the same pipe or beam. It will leave marks caused by dirt or grease on feet or fur.

FOR PERMANENT CONTROL MEASURES

I.   Try to maintain permanent freedom rather than resort to temporary reduction of rat population by periodic drives employing trapping, poisoning or fumigation.

II. After carrying out all rat stoppage measures, a reliable employee should inspect entire premises weekly to insure cleaning and upkeep, and to repair any temporary breakages in windows and doors to the outside. Allow no accumulations of rubbish to form. Watch sky-lights, air shafts, dumbwaiter and elevator shafts, and all other means of ingress from outside, for breaks. Immediate repairs to be made to eliminate openings and harborages, with rat-proof material (impervious to gnawing). Relocate or alter fixtures with hollow enclosures. Prevent careless employees or workmen from leaving lower windows or cellar doors overnight or on weekends to rodent ingress from outside sources.

III. Include in the specifications for all new construction and repair contracts a specific provision that work is be done is to leave the building in a ratproof condition. Specifications may read as follows:

> *"This building is planned and detailed, and it is the intent of these specifications, to provide a structure that will prevent the penetration by rodent vermin of any vacant space where they might find a harborage. The contractor will he held responsible for securing this condition by the closing of all points of access to such spaces, including the passage of piping and conduits through all walls, partitions, ceilings and furred off spaces, the closing of access to void in hollow tile blocks, etc. There shall be a special inspection of the building with regard to this matter before final acceptance."*

IV. All permanent measures are aimed at eliminating the rodent's food supply and shelter. Architects need to be made more cognizant of conditions that prevent rodent harborage and infestation, so as to change design of new buildings to eliminate unnecessary enclosed spaces. Rat-proof construction should receive greater prominence in the future.

INSECT CONTROL

Proper restaurant or food plant sanitation must obviously include measures for the elimination of insects and vermin. Where insects infest food, such food cannot be used for human consumption and must be destroyed, entailing an economic loss. Even more important is the fact that insects carry millions of bacteria in and on their bodies and contaminate the food on which they crawl. Food establishments are particularly susceptible to insect infestation are attracted by the food.

GENERAL ENTOMOLOGY

Life Cycle: Insects have a life cycle that consists of: (a) the egg, (b) the larva (worm-like), (c) the pupa (web, cocoon) and (d) the adult. Some insects have only three stages: (a) egg, (b) nymph, and (c) adult.

Structure: The general external structure of insects consists of three parts:

1. Head

    (a) Compound eyes - often more efficient than human eyes; enables them to see in all directions.
    (b) Antennae - hair-like structures that act as feelers.
    (c) Mouth - may be constructed for use as a hypodermic needle (mosquito), for chewing (beetle) or sucking (fly).

2. Thorax

    (a) Wings
    (b) Spiracles (breathing vents)
    (c) Feet

3. Abdomen - used for digestion, reproduction, and excretion. Insect "specks" may be from excreta or vomit.

Flies are vicious public health enemies. They breed in filth and carry diarrheal diseases to the food we eat. Flies have a life span of about four weeks. One female may lay 150 eggs at one time and usually prefers manure, garbage, or some other filthy waste matter for this purpose. In four weeks, the eggs hatch in eight hours into the larva stage (maggots) which become fully developed in four to five days. Each larva then becomes a pupa and five days later the adult fly emerges. Hence there are ten days from the egg to adult stage but this elapsed time varies according to the temperature. The female may start laying eggs four days after reaching the adult stage.

Flies generally die in the early autumn, partly due to a fungus disease that attacks them, and partly due to the cold temperature. They do not hibernate, but larvae may survive in manure piles or beneath the soil over the winter months and start the new generations of flies the following summer.

The fly cannot eat solid food. It alights on food and vomits a liquid through its proboscis-like mouth. The dissolved food is then sucked back, but, of course, all or a portion of this "vomit" may remain on the food. The excreta also remains on the food.

Examination of flies has disclosed as many as 28,000,000 bacteria on the inside and 5,000,000 on the outside of the body of the fly. Flies infest manure piles, privies, toilets, and other waste materials and then may alight on food. It is obvious that they may easily pick up millions of harmful bacteria and infect our food. They constitute a dangerous channel of disease transmission that must be eliminated.

A very large outbreak of bacillary dysentery in an army encampment was definitely traced to flies as the vectors of this disease. Thousands of cases of illnesses are undoubtedly due to flies, even though it may be difficult to prove.

Prevention: Remove all breeding places that may be found in or adjacent to the establishment, e.g., piles of manure, garbage, filth in general. Keep garbage cans tightly covered, and thoroughly clean them when emptied. Screen all openings to the outer air.

Destruction: Swat the fly, use fly paper, or use fly sprays.

Fly Sprays: These generally contain a contact poison, such as pyrethrum, as the active ingredient, dissolved in a nonodorous organic solvent, usually kerosene which has been deodorized or to which some perfume has been added. Such sprays act on the nervous system of the fly and are very effective, provided the spray contacts the insect.

Pyrethrum: Is a powder that is obtained from the chrysanthemum flower. The active ingredients are known as pyrethrins and a good insect spray should contain 100-120 milligrams of pyrethrins per 100 cubic centimeters of liquid.

Pyrethrum is not toxic to man but kills the fly by contact. When a dose of this insecticide is absorbed, it stupefies the insect by acting on the nervous system and causes the insect to drop to the floor. It may remain alive for several hours (even twenty-four hours) before death occurs, depending on the lethal dose of insecticide that is has absorbed. Pyrethrum is relatively unstable and loses strength on exposure to air or heat.

Pyrethrum Synergists: An insecticidal synergist is a chemical which, when combined with an insecticide, increases the killing power of the combination beyond what might be expected by the simple addition of each. Common pyrethrum synergists and piperonyl butoxide, sulfoxide, and MFK-264.

Golden Malrin: This effective fly-killer is a sugar bait.

Residual Fly Sprays:
Diazinon: 5-1%
Malathion: 2%
Rotenone Baygon: 1%

Note: Chlordane, DDT, and Lindane are banned in New York State. Parathion is a dangerous organic phosphate and under no conditions may it be used in home or food establishments.

INSECT CONTROL: ROACHES

The elimination of roaches in food establishments is a greater problem than flies. Roaches also carry disease bacteria on their bodies and deposit them on the food through their excreta, vomit, and bodily contact.

The most common household roach is the German roach. It is one-half to one inch in size, light brown in color with two dark stripes on the back, very agile, and very prolific. The female lays twenty-five to thirty eggs at one time, enclosed in a leathery pouch one-quarter inch long. The life span of a roach is one to one and a half years, and the female will lay five batches of eggs in one year. The egg hatches into a nymph and the nymph grows by molting its shell-like skin. After five molts, the roach reaches adult size. The length of time required for hatching eggs depends on the temperature, and ranges from one to five months.

Roaches are cunning and quick to sense danger. They can survive for long periods under unfavorable conditions and their flat oval shapes facilitate hiding in cracks and crevices into which they can glide with lightning rapidity. They generally travel and feed at night. While they prefer starchy foods, they can feed on leather, felt, or wood. All loose foods should be kept in covered containers.

CONTROL

I. It is difficult to prevent the invasion of a food establishment with insects, especially roaches, thay may come from an adjoining building or in packages delivered to the plant. The emphasis must be placed on eliminating of harborages and breeding places within the establishment as well as extermination.

II. All cracks and holes in the floor, walls, and ceilings should be eliminated as far as possible by filling with cement, plaster, putty, or plastic wood. Seams in fixtures and equipment should receive the same treatment. III. Equipment and fixtures should be placed flush against the wall and floor; if not, then a sufficient distance away from the wall and above the floor to facilitate cleaning around it. Wherever possible, wooden fixtures should be replaced with metal.

IV. All potential insect breeding places, such as rubbish, debris and stagnant water, should be eliminated. Garbage should be kept in tightly covered metal cans, and the cans should be thoroughly cleaned after being emptied. The room in which garbage is kept, prior to removal, should be constructed of impervious washable material, preferably cement, and should have can washing facilities. If this room can be refrigerated, the cold temperature will prevent insects from breeding, and odors from decomposing garbage will be inhibited.

V. Good housekeeping is a very important factor in insect control. The food establishment and equipment therein should be completely cleaned each night before closing, not only for good sanitation, but to remove all grease, food encrustation, and food particles on which the insects can feed.

VI. In addition, roaches can be destroyed with effective insecticides.

DESTRUCTION BY POISON (EXTERMINATION)

I. Space sprays are not very effective against roaches. The spray or powder used must actually contact the body and so aerosol types of sprays are not used for roaches. Where powder is used, it is generally spread with a mechanical device or "gun".

II. Insecticides may be classified as contact poisons or stomach poisons. A contact poison is one that only requires contact with the preparation to injure or destroy the insect. This occurs when the insecticide enters the external breathing apparatus or is deposited upon other vulnerable portions of the body. A stomach poison is one which is required to enter the insect's digestive apparatus in order to kill the pest. This may occur when poison is mixed with the food bait so that it is eaten by the insect, or when the insect walks through an insecticide (which may be a contact or stomach poison) and grooms itself, (licks its feet and antennae,) inadvertently ingesting the poison.

III. In extermination by insecticides, it is important to remember that the Health Code prohibits the use of any poisonous insecticides in a food establishment. Artificially colored blue fluoride powder is the only exception to this rule. When new insecticides are offered on the market for use in food establishments, they must be accepted by the Fumigant Board of the Health Department as to their non-poisonous character.

Insecticides commonly used for roach destruction are:

I. <u>Pyrethrum:</u> Non-poisonous to man but effective against roaches. It is non-stable and loses its strength rapidly when exposed to air, moisture, and sun. It is used in several forms, usually as a powder or spray. It is also used in the form of a pyrethrum vapor that penetrates every crack and crevice in the room. Various devices exist in which steam is generated that passes through a narrow horizontal nozzle to which is attached a bottle of pyrethrum dissolved in kerosene. The stream of steam causes this solution of pyrethrum to vaporize throughout the room in which the device is used, and the roaches leave their hiding places to get air and generally die on the floor where they can be swept up.

II. <u>Rotenone:</u> Is non-poisonous to man. It is a product extracted from plants known as Derris Root, Cube, and Timbo. It is an effective roach killer and is generally spread as a powder, although it may also be used in sprays.

SUSPECTED ROACH INFESTATION SUGGESTIONS FOR INSPECTION

1. Locations where commonly found Kitchen, vegetable preparation area, serving counter, garbage area, walls and shelving adjacent thereto; behind hanging pictures and signs, beneath paper shelf lining, at electric and steam meters. All warm, slightly moist, seldom disturbed dark spots.

2. Structural Defects Harborages as cracks, crevices, seams, fissures in walls, fixtures, furniture, pipes, and wire passageways.

3. Storage Conditions

   A. Accumulations of seldom disturbed paper goods as menus, newspapers, paper stock, bags, cups, cartons, containers, straws.
   B. Accumulations of empty unrinsed food containers as milk, beer and soda bottles, ice cream containers, milk cans, meat, fish and bakery boxes.

4. Insanitary Conditions

   A. Unclean garbage can exterior surface and adjacent floor area.
   B. Dried food scraps on surfaces of floor, walls, shelving; beneath and adherent to fixtures and equipment such as mixers, slicers, dicers, peeler impact about can opener and within seams of cutting boards, chopping blocks and work tables.

PESTICIDE USE IN FOOD SERVICE OPERATIONS

ARE YOU HARMING.....

.....yourself?

.....your employees?

.....your customers?

You may be - without realizing it - if you are using pestices improperly.

You may also be breaking the law.

THE LAW

State law requires that anyone using pesticides, except farmers or homeowners on their own property, be certified as a "commercial applicator" with the State Department of Environmental Conservation (DEC). This applies even to arerosol cans of pesticides, available at most grocery stores.

Persons applying pesticides in restaurants, institutional kitchens or other food service operations need special certification from DEC's Bureau of Pesticides Management. Under the State Sanitary Code, a food service operation can be closed down by the State Health Department if pesticides are applied by a person not certified to do so, or if pesticides or other toxic substances are improperly used, stored or labeled.

WHY?

All pesticides are toxic. After all, their purpose is to kill roaches, rats and other pest. Proper use of pesticides is essential to everyone's safety.

If pesticides are used improperly, food, utensils or other food-preparation equipment can become contaminated. Improper handling of pesticides can also lead to direct exposure through the nose, mouth or skin. The result may be future health problems for those exposed.

State or local health department sanitarians check for proper labeling, storage and use of pesticides during their routine inspections of food service establishments. Sanitarians also make sure that only registered businesses or certified persons are applying pesticides. Violations found during these inspections can result in fines, closure or both.

DO'S AND DON'TS

Help insure the safest control of pests in your food service operation by following these do's and don'ts:

DO be concerned about your health and safety, and that of your employees and customers.

DO ask your pest controller for proof of certification by DEC'S Bureau of Pesticides Management.

DO report uncertified pest controllers to a DEC pesticide inspector at the DEC pesticide inspector at the DEC Regional Office.

DO make a pest's life difficult by maintaining extra-clean conditions and by eliminating possible pest entry routes.

DO know what pest species you are dealing with, and what pesticides are being used. Ask your pest controller or health department sanitarian about options for pest control, including chemical and non-chemical methods. (A combination of both is often most effective.)

DO get instructions from both the pest controller and a sanitarian on what you should do following treatment.

DO read and follow the label instructions carefully when using any pesticide product.

DON'T hire an uncertified pest controller.

DON'T apply any pesticide in a food service operation yourself unless you are certified.

DON'T permit the application of pesticides while food is being prepared or served, or in an area where utensils, unprotected food or containers are stored.

FOR MORE INFORMATION .....

To learn more about pest control, or for details on how to become certified to apply pesticides, contact your local health department or DEC regional pesticide office.

RODENT AND INSECT CONTROL

HOW CAN WE CONTROL RATS AND MICE?

I. GET RID OF THEIR NESTING PLACES
    Clean up all piles of rubbish, inside and outside the premises.

II. BUILD THEM OUT
    Block all possible rat entrances. Rat-proof foundations.
III. STARVE THEM OUT
    Protect food at night. Keep garbage containers closed. Do a thorough clean-up job.
IV. KILL THEM
    Use traps for temporary control.

HOW CAN WE CONTROL FLIES?

I. GET RID OF THEIR BREEDING PLACES
    Control the sources.
II. KEEP THEM OUT
    Screen doors and windows properly. See that all doors open out and are self-closing. Install overhead fly fans or air curtains.
III. DO A GOOD JOB OF HOUSEKEEPING
    Keep foods covered. Keep garbage containers sealed. Remove food accumulations promptly.
IV. KILL THEM
    Use a pyrethrin insect spray inside the buildings.CAUTION: Don't use sprays with any food or food surfaces exposed in the room.

HOW CAN WE CONTROL COCKROACHES AND OTHER INSECTS?

I. BE ALERT TO FIRST SIGNS OF INFESTATION
    Destroy infested foods.
II. DO A GOOD JOB OF HOUSEKEEPING AND STORAGE
III. USE PROPER INSECTICIDES (CAREFULLY)

THEN IF RATS, MICE, FLIES, COCKROACHES OR OTHER INSECTS STILL INFEST YOUR ESTABLISHMENT, CONTACT YOUR LOCAL HEALTH DEPARTMENT!

ANSWER SHEET

TEST NO. _____ PART _____ TITLE OF POSITION _____

(AS GIVEN IN EXAMINATION ANNOUNCEMENT · INCLUDE OPTION, IF ANY)

PLACE OF EXAMINATION _____ DATE _____

(CITY OR TOWN)   (STATE)

RATING

USE THE SPECIAL PENCIL. MAKE GLOSSY BLACK MARKS.

Make only ONE mark for each answer. Additional and stray marks may be
counted as mistakes. In making corrections, erase errors COMPLETELY.

ANSWER SHEET

TEST NO. _____ PART _____ TITLE OF POSITION _____
(AS GIVEN IN EXAMINATION ANNOUNCEMENT - INCLUDE OPTION, IF ANY)

PLACE OF EXAMINATION _____ DATE _____
(CITY OR TOWN) (STATE)

RATING

USE THE SPECIAL PENCIL. MAKE GLOSSY BLACK MARKS.

	A	B	C	D	E		A	B	C	D	E		A	B	C	D	E		A	B	C	D	E		A	B	C	D	E
1						26						51						76						101					
2						27						52						77						102					
3						28						53						78						103					
4						29						54						79						104					
5						30						55						80						105					
6						31						56						81						106					
7						32						57						82						107					
8						33						58						83						108					
9						34						59						84						109					
10						35						60						85						110					

Make only ONE mark for each answer. Additional and stray marks may be
counted as mistakes. In making corrections, erase errors COMPLETELY.

	A	B	C	D	E		A	B	C	D	E		A	B	C	D	E		A	B	C	D	E		A	B	C	D	E
11						36						61						86						111					
12						37						62						87						112					
13						38						63						88						113					
14						39						64						89						114					
15						40						65						90						115					
16						41						66						91						116					
17						42						67						92						117					
18						43						68						93						118					
19						44						69						94						119					
20						45						70						95						120					
21						46						71						96						121					
22						47						72						97						122					
23						48						73						98						123					
24						49						74						99						124					
25						50						75						100						125					